From your friend
Kathleen Mc Nally
april / 1992

WIND IN THE ASHTREE

Jeanine McMullen

UNWIN
PAPERBACKS

LONDON SYDNEY WELLINGTON

First published in Great Britain by the Trade Division of Unwin Hyman Limited, 1988
First published in Unwin Paperbacks, 1989

Unwin Hyman Limited, 15–17 Broadwick Street, London W1V 1FP

Allen & Unwin Australia Pty Ltd
8 Napier Street, North Sydney, NSW 2060, Australia

Allen & Unwin New Zealand Pty Ltd with the Port Nicholson Press
Compusales Building, 75 Ghuznee Street, Wellington, New Zealand

British Library Cataloguing in Publication Data

McMullen, Jeanine
  Wind in the ash tree.
1. Wales. Agricultural industries. Smallholdings.
Self-sufficiency, – Personal observations
I. Title
630′.92′4
ISBN 0-04-440421-2

Set in 11 on 13½ point Garamond
and printed in Great Britain

FOR MY MOTHER
– like everything else

Chapter 1

There is a snow light in the house, making the blue and white plates
on the stone walls sharp and clear and the sea-green curtains at the
window translucent. Everything inside is swimming in light, a
bubble of bright colour against the white outside; sounds are
deadened so that the ear creaks against the silence. Mrs P swans
down the stairs in her blizzard outfit; jeans billowing out over the
tops of her moon boots, a long, bilious-yellow scarf appearing in
surprising places about her person, her Harrods hat living a life of its
own on top of a hasty coiffeur which didn't quite work, the lining of
an old coat of mine doing duty as a warm jerkin and her chihuahua,
Winston, sternly gazing at the intrusive light from under her arm.

The rest of the dogs and I have been up for hours, the dogs to step
at first delicately and apprehensively through the snow before going
deliriously wild, spinning and dancing on their long legs and then
digging up any old bones to make sure they were safe; I to look with

relief at the silent landscape and thank heaven it hasn't sent any wind to drive the snow up into drifts. The forecast had said there would be wind and last night every barn had been fortified against the blizzard, every animal on the place brought in under cover and given a bonanza of extra food and water in case I'd had to dig through to them. Wood was stacked up till there was no more room for it in the house, fighting for space with the buckets of coal, the shovels, the lengths of rope (for getting out of windows), the ladders, and the sheer panic and terror with which I fill every room when I hear the first gust of wind and see the first flake. And I had sat up waiting for most of the night, listening and opening curtains to look outside and see nothing but a lazy half moon and a sky full of stars. The snow must have come sometime before dawn, gently and without menace.

My mother cocks an eyebrow airily at me as she picks her way past to the kitchen.

'You see? I told you it would be all right,' she announces. 'No faith, that's your trouble!'

I don't bother to argue. Mrs P's faith in her prayers (enough to convince her she can stop blizzards) is not something you argue with. And there's not much any more that she can say to stop me turning into this haunted creature permanently listening to weather forecasts, testing wind directions and always on red alert even when there's nothing more desperate than a bit of hill fog threatening. It's not just the winter now that brings on this weather paranoia; the violent electrical storms which turn the hills into raging furies, as the thunder echoes and booms around them and the sharp forks of lightning hit the fields above the house with deadly accuracy, have taken a lot of the joy of summer away.

Mind you, I blame this neurosis of mine on Mrs P. It was the winter that she took herself and her salty common sense off to Australia when it all began. As if it was not enough to face blizzards and floods alone, I had the added nerve-pull of worrying about her being let loose upon the travelling public.

I always feel like hanging a notice round my mother's neck when she's about to travel, as a warning to other passengers. It would read, 'Are you really sure that this is the only boat/plane/train you can take?'; because there is no way that boat, plane or train is going to have a straight, untroubled run if Mrs P is on board. It's a miracle

12

she's never been hijacked, although I pity the would-be hijackers from the bottom of my heart if they ever trawled her in their net.

It's not that she does anything herself, just the mere fact of her being on board puts any vehicle she patronizes in jeopardy. Except that I do have a vivid memory of her once opening the door of a car, which was doing seventy up a straight road in the Outback, causing the driver to swerve into a sand patch and turn the car over so that my memory is blurred by a rain of chocolates and oranges and groceries and baby brother, as we returned from a monthly journey to town to stock up.

My nerves, haunted by such vivid scenes, are not up to driving Mrs P long distances so, when she travelled back to Australia that winter, I arranged for a friend to take her to Heathrow to catch her plane. I saw them off, content in the knowledge that our friend was a very experienced driver, indeed had at one time done a spot of lorry driving to the Continent to subsidize his farm, and returned to the cottage to indulge myself in a quiet weep over her going and try somehow to fill the emptiness she would leave behind.

It was as well I hadn't noticed that our friend was carrying trade plates on his car for some reason, so that the peaceful journey I'd planned for my mother was interrupted by the police chasing the car up the motorway and demanding to know why there was a passenger in a car with trade plates, and Mrs P was seriously keeping an eye out for a taxi in the middle of the M4. I didn't hear about it until long afterwards when the dreadful vigil by the phone was over and the inevitable 'long delay' was resolved and my brother had rung to say our parent had been duly collected and put to bed in her native land.

I believe they had one of the wettest summers on record in New South Wales that year. They even had a bit of snow.

Back in Old South Wales we had snow too; the kind that screams home in the arms of a vicious easterly wind and piles itself up against doors and windows and finds its way under the roof slates and fills the yard with drifts like petrified waves and snipes in through the cracks round the front door frame and lies in seeping piles in the tiny hallway so that there is no escape from it within or without. Alone, without Mrs P's confidence in the Almighty, the dogs and I huddled on the big sofa-bed before the fire, by candlelight for the electricity had bipped off at the first rattle of the window.

The dogs were delighted at this sudden change of sleeping arrangements as I clung to their company rather than lie upstairs listening to the slates sliding off the roof. I heard the outside carriage-light come crashing to its doom past the window and the fire spitting as the snow came hurtling down the chimney.

It raged for three days that blizzard, at least the wind kept up for that long. During the day the snow stopped coming out of the sky, but there was plenty of it lying around for the wind to play with. At night it had the real thing to start throwing about again. I longed to stay inside away from it but, because of the stock, I simply had to bury my fear and go out into the terrifying world beyond the door and found that, once I'd got there, the blizzard seemed less alien and vindictive.

When it finally stopped howling and raging and blinding, there was just the isolation of snow-bound fields and roads and an unexpected feeling of security against the world away over there. Nothing and no one could get through and, with bare survival the only priority, there was a relief from all the worries and tensions about things I should have been doing, like mucking out or answering urgent letters and paying bills or rushing off to town. I didn't even have to worry about searching for buried sheep, like so many of my neighbours, because I'd turned the old open-sided hay shed into a deep shelter and they sat warm and comfortable on thick beds of straw; sodden but safe.

It was a nuisance not having the outside light, but the snow created its own brilliance so that later, when the thaw came, the dark seemed deeper and almost tangible. By then the light had been put back and the slates were no longer piled up like playing cards on the yard and new, paler ones were disfiguring the symmetry of the barn roof.

My next visitors, after the electrician and the builder, were Bertie and Sara Ellis bearing that other lack in my life, a bottle of whisky. This was partly to celebrate the thaw and partly to oil the long interview which Bertie and I had arranged to record about some of the hazards and wonders of being a vet in rural Wales. We'd thought to have a bit of a trial run before actually going out 'on the country'. Sara had come along as a kind of memory bank and to act as a brake on Bertie's extravagant language, in case he got himself struck off. It was just as well she had come, because Bertie, still tensed up from days and nights trying to get through to his clients in spite of the drifts, and I, like a puppy off the chain after my long isolation, attacked the whisky and got sillier and sillier. Sara, heavily pregnant, stuck to her coffee and her senses. We did try a few words on tape, Bertie and I, but every time I waved the microphone about, both of us collapsed in hysterics and decided firmly that what we needed to calm us down was another whisky.

It was long after midnight when Sara cast a cynical look at us and decided we'd wasted quite enough time and tape and that somehow she'd have to fit herself behind the steering wheel of their Mini and drive her husband home. She made it, just, and then reversed straight back onto a part of the yard which my sober self would have remembered to warn her about instead of cheerfully waving her on. It looks solid enough that bit of the yard, but unless we've had weeks of drought it's a deep, treacherous quagmire under its covering of innocent-looking grass.

After half an hour of heaving and pushing, Bertie and I discovered that you get very sober standing ankle deep in slush, covered from head to toe in thick mud, so that only the whites of our eyes gleamed out, at one o'clock on a cold winter's morning. I also got very depressed when I looked at the hitherto virgin white of the cow-shed wall and saw that now it was a nasty slimy brown. The Mini didn't look too spruce either but at least it was standing clear and Sara and her bump were still intact.

'Don't stay out here in the cold' she told me. 'Bertie can wash up under the tap there. God! You both look funny!' and she collapsed in a heap over the steering wheel, laughing properly for the first time that night.

It wasn't funny though, not a bit, because somehow when we'd left the house, someone had managed to lock me out.

My screams of rage and fright brought both Sara and Bertie hurrying back up the yard. Flakes of snow had begun to drift gently down again, inside the house the lights shone and the fire gleamed and the dogs snored, but outside the stars had gone, the snow suddenly began to dance more urgently and my front door remained very definitely shut.

Sara and I turned to Bertie for inspiration. He gazed thoughtfully at the lock on the door, shook his head and announced, 'The window. Have to get in that way.'

I looked at him in disgust. If the door was locked, the windows were even more so and, even if they hadn't been, in winter they warp so badly that they'll only open a crack anyway. I've never had the luxury of being able to fling them wide, up or down. They are also very small windows.

Bertie peered at the nearest one, squinting through the tiny panes. 'That catch will give' he said and clicked his fingers at us to find him something thin and small. Hopefully we handed him bits of wire and sticks and at last the catch shifted. We all heaved but the bottom of the window remained rigid and the top only budged its usual few inches.

I was just beginning to dissolve into despair again when Bertie nonchalently leapt up onto the narrow, worm-eaten sill and began to wriggle through the gap.

'Push!' he commanded sternly and Sara and I, her bump getting in the way and my arms already trembling from the efforts with the Mini and by now freezing in their mud pack, feebly thrust at his behind.

As if on a signal, the slumbering dogs realized that someone was trying to get in through their window. As one dog they sprang to the side of the room, bounding up and down and shrieking with rage. Bertie hesitated and stuck fast. There he stayed, trapped by his midriff half-way in and half-way out, his face furiously abused by the dogs and his behind battered by two women, cackling hysterically and finally unable to do more than stand there in the falling snow with the tears running down their cheeks.

'Burglars!' cried Bertie to the dogs, flapped his arms gaily at them and then collapsed like a rag puppet, hooting and bellowing with laughter.

The dogs, maddened and appalled, raced around the room, leapt on the spot like yo-yos and then yelped in terror as the puppet came to life, gave a final surge and fell in a heap right on top of them.

There was a moment of total confusion while the dogs got themselves and their dignities untangled and I watched in agony as one of my few pieces of Staffordshire china teetered on the edge of the table and then rocked back into place. Sara peered through the window to see if Bertie had broken himself. The two of us were suddenly aware once more of the snow and the cold and began imploring Bertie to come and open the door instead of sitting on the floor laughing as the dogs licked their very own burglar's face.

I didn't go back to the car with the Ellises this time, but watched the tail-lights of the Mini disappear through the falling snow, thoughtfully drained the last of my whisky and took a tape measure to that gap in the window. It was no more, try as I might to make it bigger, than five inches at the most. Not exactly the eye of a needle, but something very like.

Chapter 2

That was the Winter of the Children. They came, Bryony and Jason, Siân and Dafydd, knowing that publicly they were considered little angels and in reality were being little devils.

'Going to help Jeanine,' they'd solemnly inform parents, importuning them to muck out the stock or at least their own bedrooms, as they departed virtuously and arrived giggling happily to assist me in nothing more arduous than the carrying of biscuits and mugs of squash or tea. And they sat, kicking their legs up in glee as they told me the latest gossip of the school bus or listened to my old records; records that had been played to many eight and nine-year-olds over the years. Like them, even this farm-wise little bunch got carried away with Rat and Toad and Mole and Badger yelling 'A-Walloping We Will Go', or (and Dafydd, driving now on your big orange tractor or sinking a mean pint with the boys at the pub, don't blush) Mrs Tiggywinkle chortling 'There is nothing in the world that's nicer, I can happily guarantee, than a real hot, strong cup of tea.' And while their children entertained me and themselves, their parents got on with the work and felt a little relieved of any responsibility to enquire after me on their own account. For there I was, alone for the first time on the farm and heaven knew how I was coping.

In all the years since I'd first drifted casually into the valley and been caught and held by the little cluster of buildings with their guardian ash tree towering above, and found myself suddenly negotiating to buy the lot and a few surrounding acres, I had never really faced up to what living there quite alone might be like. At first there had been my friend the Artist, as bewitched as I was with the whole adventure, who had stayed to look after the farm and the animals while I lived the half-and-half life of the commuter. Finally the loneliness and the sheer grind of the day to day reality had palled on him and he'd left, but then Mrs P had ignored her own precarious health and the rest of our family 12,000 miles away, to come with me when I had to take over things myself. For the past few years she had made sure I got a hot meal, even coped with the stock when she had to and, above all, lifted me from the gloom I find it so easy to sink into when things go wrong.

For go wrong they did and for a while it had been as if the whole mad venture had begun again, but this time without the wild dreams and the crazy optimism with which the Artist and I had been blessed at first. Things had got so bad, what with the lack of broadcasting work, now that I'd moved from London, and a kind of brain paralysis which could think of nothing further than the next feed or water bucket and the imminent collapse of the buildings, that eventually it had seemed as if I would have to sell up to another dreamer. But, as so many times before, help had arrived miraculously with a legacy from my father and now, with a breathing space and enough money to buy in a bit of basic help if I needed it, Mrs P had gone home to Australia and taken with her the light and left me to beat back the shade single-handed. And so, for a while, the children came to beat it back with me, until I was getting decidedly fishy looks from their parents and began to suspect that I'd soon be labelled 'Child Exploiter' if I didn't shoo them away. I could hardly tell on them and destroy their carefully manufactured image of willing little souls mucking out Jeanine's goats and horses and carrying buckets of water for her. So the afternoon parties came to an end and for a while there was silence where there had been delirious giggling and into that silence came The Bore. And did he fill it.

He was travelling round the world and was wished on me by some not-too-distant neighbours who thought he might be able to give me a hand over the winter months. He was looking, it seemed, for somewhere to hole up until the spring. I should have wondered a bit more why they weren't more anxious to make use of him themselves. There's never too much good help on a farm and, if it comes free for board and lodging only, not too many farmers will sniff at it.

He arrived, not the big strapping lad I'd expected, but a small, slight one, barely visible under the folds of an ankle-length army greatcoat and on his feet nothing more substantial than a pair of sandshoes. A few shirts and a thin pullover were the only other gear he had to combat a Welsh winter.

'Perhaps,' I thought, 'he's really tough.' But he wasn't, just not intending to do anything much except sit by the fire while I cooked his meals and he dreamt of his travels or pulled all my books out or scribbled endlessly in a notebook. I did try to get him fascinated by the finer points of making a good fire, or the basic care of the stock but his indifference was supreme. He had a good memory though because he quoted back everything I'd told him a couple of hours later, as if he'd known it all his life and I, poor fool, needed educating.

He was most active at about 2.00 a.m. when his muse hit him with a wallop and suddenly he needed more space to scribble in and more of my books to consult and more fire to warm him and tea to revive him. Reviving him in the morning was not so easy. By midday I'd usually given up and gone to gibber with rage up on the hill with the sheep. The day I lay down on the cold ground and screamed with fury, I knew I'd had enough.

The end came, almost fatally, when my aid and helper, in a last bid to get to grips with the wilderness, decided to go for a walk to the lake, a couple of miles up the mountain. There was a thick fog; the snow lay, if not crisp and even, still pretty deep in places, and the radio was telling people to stay off the mountains. My words of warning only egging him on more and my offers of proper boots and gloves treated with scorn, he set off in the greatcoat and sandshoes outfit, a lone, brave figure against the elements.

He straggled back about an hour later, frozen stiff, wet through and thoroughly frightened. The greatcoat hung dripping over the kitchen all night and he lay in bed sneezing frantically into the hot toddies I made for him.

The next day I rang his friends and demanded that they remove their property. They seemed strangely reluctant to do so. I last saw him giving them a long and detailed account of his smallholding adventures; adventures he must have been dreaming about while I was battering on his door to get him out of bed, if only to partake of a little light lunch. It was the last time I ever accepted 'free' help.

It took me a while to get over that fortnight, visitors of any kind were headed off at the gate, but finally I gave in to a persistent request from an American lady living in one of the big English seaside towns. She was making, it seemed, a small film about women who were 'going it alone'.

Of course it was sheer vanity that made me say yes, even if it meant I'd have to put her up for the week-end. I didn't get much chance for vanity these days. My broadcasting career had receded further and further into the past and, although the legacy from my father had meant a temporary respite from all the deep financial worries I'd had for the last few years, the future still looked bleak from the work point of view. Every now and again however, the odd bit of flattery from someone who didn't know I was a has-been, helped to make me feel I still belonged to the world of studios deadlines adrenalin pumping and life lived on the instant.

I met her off the train which puffs into the little wayside station seven miles away. It's a delightful train which takes its time to potter from Swansea through Mid-Wales up to Shrewsbury and, if you want to get out at our station, you have to tell the guard to stop it for you. If you wish to catch it, you flag it down or it might just puff straight past you. I say 'puff', even though it's a diesel, because it has the gait and atmosphere of a steam train and sometimes I think that it, too, is fantasizing about the past; like my visitors who arrive by it and get out with a fatuous grin on their faces as if Ivor the Engine himself had delivered them. The helpful guard adds to the old-fashioned image and, being Welsh, you get quite a bit of pungent wit thrown in.

At one time I always seemed to be meeting someone off that train or re-packing them into it and the guards got to know the look of my particular brand of passenger.

'There you are then! All safe now,' they'd assure my latest import from London as they staggered down the steps. 'Didn't think we'd really stop!' they'd chuckle at me with a nod. But the day the Lilliputs arrived, the guard on duty was very quiet. The Lilliputs, as their friends called them, were both very small, she a dancer and he a photographer, both exquisitely made and their clothes the very latest in high camp fashion. They looked, on this occasion, with their close-cropped heads, shiny jackets, tall, pastel, leather boots and a way they had of walking in unison, the very essence of the Space Age. The guard helped them down silently and then turned to me slowly, his eyes doubtful.

'These belong to you then?' he asked dryly and, as I nodded dumbly, he stalked deliberately up the platform slamming doors with deep precision and blew his whistle in slow motion. As the train picked up speed, I saw him still gazing back in disbelief at the two visions on the station.

I wondered, therefore, what effect my Film Maker would have. I certainly expected to see her emerge loaded down with cameras and tripods and reels of film. I also expected someone quite different to the tiny lady who hopped down the steps with nothing more than an overnight bag, beaming at me enthusiastically from under a 1920s 'bang' and tripping about on little Anello and Davide, strapped, dancing pumps. I immediately made a grab for the guard, thinking that her equipment must be in his van and that he'd forget to hand it out. He was far too busy fussing over his passenger with whom he seemed on the best of terms. Both of them were lost in admiration of the way, yet again, the little train had stopped at the lonely strip of concrete hidden around the bend of willow trees.

'Cameras?' exclaimed the Film Maker when I began to panic about her absent equipment, 'why I've got it all in this bag here!' and she turned from me to wave merrily at the guard who continued to wave back as the train chugged on its way and released the cars waiting at the level crossing.

As I tucked her into my old Beetle and set off up the village street, I did worry again that it was still winter in the hills and my cottage would feel like an ice-box to an American. I'd warned her as much on the phone, but she'd assured me she was naturally centrally-heated and in any case could bear great discomfort in the cause of her art.

I'd accepted all this, but I'd really imagined someone just a bit younger and tougher and those dancing pumps didn't quite look up to wading through the mud. However, I'm always a bit inclined to take people on their own assessment and when we arrived at the farm she skimmed across the muddy yard like a small bird and was too lost in general admiration of the cottage to notice the cold at first.

The truth came out later though, when I was showing her into what was usually my mother's room and explaining carefully that the antique chair with the carved wooden skirts was really a commode. Her eyes lit up like stars, she threw her arms wide in relief and cried, 'Thank God! I'll use it! I was trying not to think about using that arctic bathroom of yours in the middle of the night!' And bless her, she emerged next morning clasping the bowl as if toting a po around was all part of the fun.

I enjoyed having her. There are some people who mar the place with their presence, making you look at it again yourself and find it wanting, everything you've loved suddenly appearing grey and mucky and of no account. But there are others who give it all a new dimension, reminding you of your own excitement when everything was new; small things you'd taken for granted seem wonderful again and the animals acquire a new value and resume those personalities which often get lost in day to day familiarity. When people like that go, you feel re-charged, like coming back from a holiday, ready for anything.

My Film Maker was the re-charging variety and when she wasn't tottering around pointing her tiny camera at me as I fed the animals or took them for walks, we sat by the fire and she told me about the dramas of being retired amongst the retired; for retired was what she officially was, she and her invalid husband. I learnt about the morning race to the library to lay claim to the papers before anyone else; of the way she picked up oddments of material and sewed them by hand into the multi-coloured dresses she wore with panache; of the way she saw Britain by making her films which were eventually shown at cinema clubs, and most of all I learnt about her terrific appetite for just being alive and enjoying every minute.

Only once did she fail to endure the rigours of the farm. I'd taken her up the steep field at the back of the house to show her the effect of the sudden rise of the land on the view, which changes from step to step. Striding ahead I turned, expecting to see her beside me exclaiming happily about the long, rolling prospect of the valley, but she was still standing a few yards from the gate, her small face bereft.

'I can't!' she quavered. 'I'm stuck.' And she had to be gently brought down to the yard again and her confidence in her ability to go anywhere and do anything, quietly restored.

My own, ever failing confidence had a rather strange boost a few weeks later when the Wizard came to see me. I call him the Wizard because the first time I saw him on another country station in Norfolk, he was wearing a voluminous caped coat and a deerstalker, and because of the magnificent horses he bred. But he also had a few ideas about the way you can twist fate to your own advantage and, as I was going through a bad patch at the time and told him that I could almost believe someone had ill-wished me, he gave me, looking around and begging me never to tell anyone where I'd got it, a spell to ward off the evil eye. I've used it too. Whether it does any good or not I can't really say, and maybe I've got bits of it wrong anyway, but when the gods are chucking mud at me, I'll try anything.

Anyway, the Wizard turned up in Wales and absolutely filled the cottage both with his size and his personality so that to get him outside was a matter of sheer survival. Like the true farmer he was, he showed a keen and lively interest in the horses, the goats, the sheep and the buildings, but then he fell curiously silent.

I have a habit of sitting down amongst my animals because I've found that if I get down to their level, they'll usually either come up to me or just quietly get on with what they're doing, instead of flying off in alarm. This is especially important if I've got someone with me. They're so used to seeing me by myself, that if there's anyone else about it's likely to be a vet or, as far as the sheep are concerned, my neighbour Myrddin coming to do something nasty like take them off for dipping.

Until the time the Wizard came to visit, I never gave it a thought, this business of sitting down amongst the animals, talking to them and even singing them little ditties.

For the first few years after I bought the farm, the Artist looked after the stock while I rushed back and forth to London to make enough money to support the mortgage and the feed bills. When he finally left and I had to take over myself completely, I was really a stranger to the animals and, frankly, some of them scared me stiff. A lot of them had grown very wild in the months of changeover and, like a lot of insecure creatures, bit or kicked first and thought about it afterwards. In order to calm them down and steady my own nerves, I took to singing that song from *Camelot*, 'How to Handle a Woman', except that I'd substitute 'horse' or 'goat' or 'goose' for 'woman'. The animals were so amazed they stood still, and the words of the song helped me to keep my blood pressure down and a smile in my voice and try to remember that loving an animal will get you where hating it won't. Even the cantankerous bunch I've collected over the years got the message, including Douglas the gander who usually waits till I've finished chortling away before he bites me.

By the time the Wizard came to visit us, the animals and I had a pretty good working relationship and, because for the past few months I'd been almost entirely alone with them and we all knew each other's idiosyncracies very well indeed, maybe there was something in what he had to say about it all.

'It's very, very odd,' he said thoughtfully, 'but when you're with those sheep, you seem to become a sheep yourself! When you're with the goats, you simply merge into their herd and become part of it, and those horses half expect you to break into a gallop with them!'

I began to protest but the Wizard held up his hand and stopped me.

'I've got a feeling,' he mused, 'that you've discovered a new way of communicating with animals. Perhaps you should do a bit of research on it or something.'

Now one person I've always envied is Dr Doolittle, but singing off-key and sitting down in the mud with them is hardly in the same class as actually speaking intelligently to the animals. The Wizard, however, was deadly serious. It took me a good deal of persuasion to stop him actually writing a paper on it.

He left earlier than he'd intended and I've felt very guilty about it ever since.

I may have got on better terms with the animals, but the cottage and the buildings were, and still are, waging a relentless war. The main key to survival here is to remember that everything is held together by string and prayers and has the delicate constitution of fine bone china, until you want to do something like knock a hole in it and then it withstands a battery of sledge hammers. Like most large men, however, the Wizard found it hard to shut a door quietly or not to lean up against a wall too heavily and after a couple of days both our nerves were a bit frayed. The crisis came when I was sitting in my narrow little kitchen which has a large black beam, seriously reducing the headroom, running straight across it. It's too low to hang things on, unless you want to keep banging your head, but just the right height for collecting all the Rayburn dust and rather too high for me to remember to dust it very frequently.

The Wizard sauntered into the kitchen to have a chat, hooked his long arms up and leant over the beam. Remembering the dust, I leapt up to warn him but he shrieked with fright, clutched his heart, staggered back, yelled 'My God! My God!' and fled from the room.

I was amazed that a little dust could have such a dramatic effect.

'I'm terribly sorry,' I said, hurrying after him, 'I meant to dust that beam this morning and I forgot. I hope you're not too filthy!'

'Dust?' He looked at me aghast. 'Dust? Woman I thought you meant the whole beam was coming down! I think!' he murmured, 'I'd better be getting back to Norfolk.' And nothing I could say would stop him.

With him went the last serious bite of winter, and spring, for the first and last time I can remember, came very early that year. The dark spears of the daffodils were already tall and full, the buds on the bushes in the garden plump and promising and over the trees down in the wood, there drifted a faint haze of coral. The earth, which until then had smelt of mud and muck or of nothing at all when the frost bit deep, began to send up that rich warm perfume which hints of a thousand hidden delights and, above all, the promise of sun.

Beguiled and forgetting that March is a treacherous time to believe the weather, I suddenly needed, like Mole when he made his bid for the upper air, to be out and doing and away from the lovely, but sometimes claustrophobic, hills which ring the valley. I resisted the impulse to leap in the car and drive to the edge of the world for the simple reason that my car was on the blink. In fact, just to get into town to do some urgent shopping, I had to ring my neighbour Myrddin Parry and beg a lift.

'Yes fine,' he said, 'but I have to go on to Tregaron tomorrow to get some tractor parts because once lambing starts I'll be stuck here.'

Tregaron, although not the edge of the world, was far enough for the moment, and any trip with Myrddin tends to be a bit of a magical mystery tour because of his passion for taking back ways through the hills or exploring barely-made forestry tracks.

Going away for more than a couple of hours is not something (even in the mood I was in) that I find easy to do if there's nobody at home. With no Mrs P to man the fort, I had hardly been out all winter. Apart from some disaster striking the place, I'm haunted by the story of a lone farmer who once left his place to go into town and had an accident. It was only days later when anyone thought to check on his stock. Most of them were locked up and by then half dead with hunger and thirst. So before I leave, I go into such a frenzy of making sure that every animal has enough food and water to keep going for ages and that they can escape if necessary, that half the time I end up not going at all. This time, however, the need to get out overruled everything. Fortunately the dogs and cats in the house were pretty well trained to go into a state of suspended animation; my two old Burmese cats would curl up together for hours and the whippets as a breed are rather like cats in that respect. The blue whippet Merlin, however, I usually took with me for the pleasure of his company.

By now he was just coming up to eight years old and, since I'd nearly lost him a year ago, he'd become even more precious. He was a unique dog in every way and it was largely due to him that I hadn't felt the lack of human company that winter. Merlin understood every word I said and often made a very fair attempt to answer me, sometimes gazing at me in despair when I failed to interpret the extraordinary range of sounds he could make. In that respect at least the Wizard had been right, although the communicating was a bit one-sided. The dog had learnt how to understand the person; the person was rather more stupid.

When Myrddin swung his car into the yard, I stood hopefully with Merlin at my side. As their names are simply English and Welsh for the same thing, Myrddin had a special affinity with Merlin to whom he invariably spoke in Welsh and was just as convinced that he could understand him as I was that he could comprehend my English.

With Merlin sitting in state on an old coat in the back of Myrddin's immaculate car, we did a lightning shop in the local town and then set off along the hidden by-ways and odd little roads, till we came to the great Llyn Brianne reservoir which winds its way round the valleys and laps at the foot of hills shrouded in forestry plantations. Spring may have been smiling prematurely on the rest of the countryside, but here everything was cold and sinister and dark.

Myrddin stopped the car and then led the way down to a small promontory of rock. Below us, visible under the still water, were the stone walls of a farm house.

'It was empty, mind, long before the water rose,' said Myrddin. 'There was a murder or something there once. But I can remember driving down past it not so long ago,' and he shook his head remembering a road where now only water stretched across the landscape.

I was suddenly thousands of miles away, seeing another great reservoir that had been created in the Snowy Mountains of Australia.

When my mother married her second husband, Arthur, he had forsaken the land and gone to work in the little town of Cooma, just when the huge hydro-electricity scheme was being bulldozed out of the Australian Alps. It was an amazing engineering feat and no one had much time or sentimentality to spare for the little settlements which were flooded when the water filled the valleys. They built bigger and better ones above the encroaching water line, but I remember hating the sight of that water creeping up, of tiny orchards being swamped and of so much work and endeavour sinking out of sight. I understood then, why there are so many legends about bells that ring from sunken cities and stories that the life which has been submerged still goes on down there; the lost Atlantis we are all still looking for.

Even Merlin was subdued as we set off again from Llyn Brianne, this time down a forestry track that had our teeth juddering, and the trees brushed the sides of the car and seemed to drown us as effectively as the water we had left behind had drowned the little farm. And then we were out again into open country; country which was wild and desolate and still streaked with odd patches of snow. The small, squat farm we passed, with its long stone wall straggling around it, had that strange remoteness about it which such places never seem to lose even in high summer, as if that summer were alien to them and they are only truly themselves when fighting off shrieking winds and deep drifts of snow.

On we went, climbing and twisting and descending in that bleak world still caught in winter, until at last Myrddin spoke again.

'That,' he said, nodding to a clump of pine trees ahead, 'is one of the most remote chapels in Wales . . . Capel Soar-y-mynydd.'

'A chapel!' I exclaimed. 'Here? Who comes to it?'

'I don't know about now,' he replied, 'but there used to be a lot of people here once. They built the chapel from the river stones.'

He stopped the car and we got out and walked to the chapel standing quite alone in its clump of trees, neat and tidy and open to all comers.

Inside there were the tall wooden pews with their doors at one end and, in front of them, a high pulpit and for one giddy moment it seemed as if the chapel was crowded with people. I shook myself and re-focused on the present where Myrddin had found a huge book of signatures which he was poring over eagerly.

'I should find some of my relatives' names here,' he said.

I left him and went back to the car to get Merlin. He danced, the light glinting on his gun-metal blue coat, amongst the gravestones and around the sighing pine trees and it was like a camera shutting and clicking in my mind as I watched him.

Two weeks later he was dead.

Chapter 3

Even now, so many years after, I find it hard to talk about; to go back over those dreadful days; those weeks and months of grieving. Indeed it was longer, but dare one admit to grieving so much for a dog? I did, and I do.

Actually in the case of Merlin I was not alone in my grief. So many other people had loved him; so many had been enchanted by his beauty and his wonderful way of 'talking', that when he died they too felt something very rare had gone.

It was a sudden bout of pneumonia which took him; pneumonia he might have recovered from if his heart hadn't been weak. I sat with him day and night till I could hardly move myself, only leaving him to dash out and feed the stock, resenting every minute away from him and even looking bitterly at the rest of the dogs.

'Why?' I moaned, 'does it have to be him? Out of all the dogs, why does it have to be the best?'

The other whippets, poor little souls, crept around, not once demanding their food or walks, or huddled on their bed out by the kitchen, grateful for any notice I took of them. I've learnt many times since then that it *is* always the one you love or value most that goes. Like the Chinese crying 'Bad Rice! Bad Rice!', when they had a good crop so the gods wouldn't be jealous, I find myself now always denigrating the best I have. I'm not the only one. Myrddin once told me that if he had a really perfect calf or the best lamb he'd bred that year, then he could be sure that was the one something would happen to. It pays to look sideways at your best animals and expect the worst for them.

For Merlin I still hoped, unable to believe that like the last time, when he'd almost died from eating the remains of one of the pigs which had been put down with a massive dose of anaesthetic, he wouldn't suddenly, miraculously recover. If he'd been an old dog perhaps I might have been able to bear the thought of him dying better, but at eight I'd expected to have many years of him yet. But I had to face the fact that it was probably his previous near encounter with death which had weakened his heart still further and, in spite of my hope, there was a dreadful sense of inevitability about it all this time.

Bertie came and did everything he could but said finally: 'Miracles do happen and sometimes, if you fight hard enough, you can achieve more than my injections will do. It's up to you and him now!'

Which was why leaving Merlin even for a moment was misery. I was trying desperately to give him my own strength to fight with, and it seemed as if I fought for every breath with him.

By the third night, however, my ability to fight off sleep was waning. I sat there by the fire in that room, hating it for the first time, feeling the walls close in, almost unable to bear the dreadful rasping in Merlin's chest and begging not to fall asleep because I knew nothing would have woken me.

It must have been getting on for nine, when a knock sounded above the howling wind outside. I didn't know who to expect at that time of night and I was past caring anyway. I stumbled over to open the door and standing there, carrying a basket and a sleeping bag, was Sue Kingerlee.

'I've come to keep an eye on Merlin so that you can have a rest,' she said quietly.

Sue and her husband Darrell had opened a bookshop in Llandovery a few years before. I shall never forget the great excitement on the day I saw their newly painted sign and the realization that once again I could have the joy of browsing through and buying books. None of us thought they'd last of course, but they did and nowadays I can read a review, order a book from the Kingerlees and have it quicker than a big city shop would get it. In fact, many times they have the book in stock anyway, long before the big boys would have given it a thought. Their shop is seldom empty.

In the beginning, however, Sue and Darrell spent a great deal of time worrying over their new venture and gazing at the empty winter streets and along the even emptier spaces in their shop. When Merlin and I came in therefore, they had plenty of time to fuss over him, and he responded by entertaining them and taking their minds off the lack of trade. It became so that if I did pop in without Merlin, they were visibly disappointed and even if I bought a book it wasn't quite the same without Himself there to help pack it up. And now Sue stood on the doorstep, understanding completely how I must be feeling and prepared to give up her own sleep to watch over the little dog who had won them so completely.

She stayed until noon the next day, helping just by being there, letting me get a few hours sleep, watching Merlin while I attended to the rest of the animals and even making me the first hot meal I'd had in days.

I'd better get back and see how my family are surviving,' said Sue as she left. 'I'll be back tonight or, if you need me, just ring and I'll come straight away.'

An hour later Merlin tried to give one of his little honks, stretched his long neck out and went limp. It took me another hour to accept that he was dead.

Outside, the bush of scarlet *ribes* was in bloom earlier than I'd ever seen it. It reminded me suddenly of the three red May trees that had been covered with blossom when I'd taken Merlin as a puppy for his first walk in Kensington Gardens. I gathered the limp body in my arms and took it outside to the vivid glory of the bush and spoke to my dog like a fool about all the lovely things we'd seen together, and then I took him inside and laid him in his favourite blanket and vowed that never, ever again would I love any animal in quite the way I'd loved him.

The next day Sue and Darrell both came to help me bury him under the old plum tree in the garden. When they left it began to rain and the wind lashed the tree. I stood, it seemed forever, at the window and gazed at the little grave. And in my mind, over and over again, like a dirge, ran those lines from Gerard Manley Hopkins' poem *Spring and Fall:*

'Margaret, are you grieving

Over Goldengrove unleaving?'

and like a dull drum beat came the final line as answer:

'It is Margaret you mourn for.'

Margaret is my second name, a family one, and when my mother planted it on me she gave me a bit of the family curse, for none of the Margarets had a happy life. And when things go wrong it's the Margaret in me who glooms her way deeper and deeper into the depths. As I was descending now, understanding at last what T. H. White had meant when he wrote to his friend, David Garnett, about the death of White's own beloved setter Brownie:

'I have found out how people "die of a broken heart". It just means that they lose interest in being alive. Also, it is not the deceased person that dies (for them) but it is themselves that die: all that they consisted of, for the last 12 years in my case, steps into the past leaving them to start a new life all over again, for which, if old, they lack the power to re-organise and re-integrate, and consequently they give it up. Brownie has been quartering in front of me for 12 years, while I have plodded behind that dancing sprite, so now it is difficult not to follow her still, into the past.'

And it was there, in the little grave out in the wild wind and rain, that so much of my past lay.

I'd found Merlin in the heady days, not long after I first broke into broadcasting, when I felt I had everything before me; when I'd first found my own little flat in a part of London I loved; when life was full of friends I valued and the Artist had seemed to be a quiet, calm part of my life like a warm, steady wall to lean against. Merlin had been with us on that first trip to the Black Mountain where we'd met Gerald and Imogen Summers and fallen in love with their way of life amongst their birds and dogs and we'd wandered around till we found the neat, whitewashed farmhouse nestled against a hill and looking out to the spectacular heights of the Carmarthen Fans.

Merlin had been with me on those frantic dashes to London to work and on the joyful returns to Wales where the Artist and his lurcher Elke waited for us.

Merlin had been there when we experienced those never-to-be-repeated days of discovery about the countryside around us, the farm itself and each addition to the stock and their various idiosyncracies; days when everything had been a mad adventure and every survival of every crisis a matter of huge triumph. Those days never come back. You may know more, be able to cope better, even be more solvent, be more accepted by your neighbours (something you once passionately longed to be), but it is gone, that crazy innocence, that delight in the smallest step forward, that despair in every failure, that vivid intensity of living. I get a lot of letters now from people who are embarking on their own places in the country and, although part of me is these days a little impatient with their assumption that they are utterly unique, another part envies them that feeling that it couldn't possibly have been like that for anyone else. A bit like being in love.

Merlin had been there leaping through my own dream days, but now the Artist had long gone, much had changed in the valley itself and on the farm, several years of struggling alone to keep it together, with nothing but a mounting pile of debts, had rubbed any stardust out of my eyes. Perhaps the one last speck of it had been Merlin himself; a speck which gleamed back through a thousand happy memories. Now he was gone and those memories rose up and swamped me till it was almost unbearable. When finally they left me I knew that I was really alone with only the present to make the best of. Except that, next morning, a very odd thing happened.

My two Burmese cats, Suyin and Pip, always slept on my bed and every morning Merlin had trotted up the stairs to join them. He'd never sleep upstairs, just come to spend half an hour with us before I finally got up. Sometimes, however, he didn't jump up on the bed, but simply sat in the doorway waiting for us. On the morning after he died, after a night of very little sleep, I got up drearily and the cats jumped down to stalk before me across the room. Both of them checked at the door as if there was an invisible barrier there. Both of them yowled in irritation and tried to pass again. Suyin, the older and more philosophic, jumped over whatever it was and went on her way downstairs grumbling, but the little cream cat, Pip, who was never very bright, tried twice more to walk through the door before suddenly, the barrier was gone and she could go on her way. That night and for a very long time afterwards, when I opened the door for the dogs to come in after their last run, I waited and counted an extra one in.

Two days later Dolores, who in goat form was another Merlin in character and beauty, nearly died giving birth to two dead kids and a small, weak, live male. The dead ones had been mouldering away inside her for a day or so and her own life was touch and go. Like most goats, her will to survive vanished as soon as she felt bad and so it was another battle of wills to badger her into life again. To encourage her I didn't have the male kid put down (which is the kindest thing to do if it isn't good enough for a stud goat and you don't want to rear them for meat) and finally she deigned to live.

I've learnt to hate that phrase 'Things can't get any worse'. In fact I have been known to get a bit violent when people blandly trot it out. Things have an enormous capacity for getting much worse and sure enough, as soon as Dolores was off the danger list, four of the sheep, heavily in lamb, decided it was their turn, flatly refused to eat and just sat about intent on dying.

It was Twin Lamb Disease, or Pregnancy Toxaemia to give it its right name. Very basically, it means they haven't been getting enough food to cope with a heavy lamb burden, the ewe uses up all her sugar reserves and that in turn causes other problems, which are often fatal, unless you can build up the blood sugar content fast.

My small flock of Black Welsh Mountain sheep had been getting enough food all right but, distracted as I'd been, I hadn't supervised them to make sure that the big bullies didn't get most of it. The four sick ones had just been a bit too slow to get their fair share, so now it was back to dosing and cajoling and all-night stands again. I saved them by giving them the 'Leo-Soup', a concoction I dreamt up consisting of a marvellous cattle restorative called Leo-Cud mixed with lashings of honey and glucose. They got it on the hour, every hour, whether they would or no and finally gave up their unequal battle to die and popped their lambs out as good as gold, along with the rest of the flock.

I thought that lambing, which was now in full swing, would take my own last reserves of energy and was seriously thinking about having a swig of the 'Leo-Soup' myself, when the inside lot began to fall by the wayside. First there was little Pip who developed some mysterious blockage which nearly killed her, and then one of the whippets, a fat little soul called Boy, who'd been devoted to Merlin and gone on fretting for him, grew weak and wan and very ill.

Bertie came and collected Boy. He'd practically lived on the place anyway for the past few weeks.

'Why?' I demanded as he was leaving with Boy, 'Why is so much happening to me? I don't neglect these animals. You know I fuss over them like an old hen. There's even a bloody sparrowhawk after the poor old pigeons! All I need is for the horses or Blossom the pig to start getting sick and I'll give up!'

'It's the Blight,' said Bertie grimly. 'Nothing to do with neglect or anything, just the Blight! It's hitting another client too . . . when I'm not here, I'm over at his place. Everything's just keeling over for no good reason and he's quietly going off his rocker. Often happens. I don't hear from someone for ages and then suddenly they start ringing up about one thing after another and I'm over there night and day till the Blight leaves them and starts on someone else. Mind you,' he grinned, 'it's very good for business!'

He looked at me thoughtfully as I tried to smile back. I must have been a revolting sight by then; sleepless for weeks, subsisting on hasty cups of tea and bread and butter, my clothes creased and smeared with whatever goo I'd been dosing something with, and my eyes blood-shot and wild from gazing at one horror after another.

'Sara and I have been talking things over actually,' he said at last. 'We think you need to get away from here for a bit . . . you get too emotionally involved in it all. If you can get someone to come in and feed the poultry and the sheep, we'll have the horses and goats and the dogs over with us.' He coughed and looked away, embarrassed, as the tears started rolling down my cheeks. I can just about stand the horrors; what really breaks me up is kindness.

'Think about it,' Bertie commanded as he drove away. 'And for God's sake make yourself a decent meal! You look terrible!'

I did think about it, but somehow the effort of making a move was too much. It was easier just to grind around where I was and, by the time Bertie returned with Boy, healthy and bright-eyed once more, the spring had really come to stay. Dolores' little kid was whizzing around tormenting the other goats, the lambs were beginning to choose their gangs for racing up and down the field, the birds were collecting Doli's and Gwylan's discarded horsehair for nests, the geese and ducks were mating as if their lives depended upon it down on the pond and the chickens were sneaking off furtively into the nettles to their secret hoards of eggs. Somehow I'd got through that winter with no Mrs P to bolster up my confidence or sympathize or make me warming meals and keep the fires going. And the Blight, for the moment, had passed on.

I was trying to do something constructive about all the accumulated mess and muck that had piled up while I was too busy trying to keep everything alive to deal with it, when the phone rang shrilly across the yard. I ignored it for a few rings, but it went on and on. It's always a problem the phone. Better, I sometimes think, not to have an outside bell and then you don't have the dilemma of whether to rush in and answer it or go on with what you're doing. Nine times out of ten, by the time you've parked your milking bucket or muck fork or whatever and slithered and slid across the yard and panted up the steps (trying to run now and muttering 'Don't hang up!'), it stops, just as you reach for the receiver. Then you have to cope with your blood pressure and stomp outside again and wonder what it was you were doing and curse whoever it was.

This time they didn't hang up however and, as I puffed 'Hullo', there was that distant, echoing click which means the call is coming from a long way away.

'It's me!' cried Mrs P's voice, 'I'll be coming back in ten days. It's hardly stopped raining over here. What's the weather like there?'

'We've had a lot of rain here too,' I said. 'But the sun's just come back and it's shining like mad all over the place!'

Chapter 4

It was shining in that sharp, but rather intimate way it has in spring, so that instead of glaring off everything as it does in high summer, it deepened the greens of the new grass and put clear tones into the gold of the daffodils which were clustering all over the garden or popping up in surprising places under bushes or on the banks; legacies of past gardeners and loved best of all because I never quite knew where to expect them. The barns were milky-white, streaked with shadows, and the soaring cliffs of the Fans navy blue against a pale sky.

The goats took themselves off to the river meadow and came back hours later utterly content with the rich variety of things they'd found to nibble; chickens and ducks preened and groomed themselves till every feather gleamed; Blossom and the two horses rolled deliriously and then sat soaking the sun into their bones; the sheep snatched at the grass as if they'd forgotten what it tasted like, and their lambs finally took time off from racketing about to do their annual job of flattening the molehills for me.

It's by far the best way to get rid of the wretched things because if there's anywhere a lamb loves to lie, it's on a fresh molehill and the great advantage of that, over trying to swipe them down yourself, is that the stones, which the moles heave up along with the earth, don't get flung all over the field. The lambs simply press the whole lot back where they belong. Even ancient molehills, which have been grassed over, have their uses where lambs are concerned for, as that wise old husbandman Thomas Tusser said:

'If pasture by nature is given to be wet,
then bare with the mowlhill, though thick it be set.
That lambe may sit on it, and so to sit drie,
or else to lie by it, the warmer to lie.'

Which little ditty I repeat to myself when I go down on to the meadow and see those remorseless, grass-covered mounds everywhere, long past anything but a bulldozer to flatten them, and see my neighbour's field across the river, flat and green and perfect. I've always thought it odd therefore that, until we put up a fence on his side of the river, his ewes spent a lot of their time on my weedy meadow. Probably looking for molehills to park their lambs on.

The sun stayed with us, ironing out the kinks of winter, until two days before Mrs P was due home. Then the rain came back, gently at first but finally bucketing down and driving the sheep back to their hayshed, the chickens to revert indignantly to their perches in the poultry house, Blossom to huddle in her home-made bed of bracken in her shelter and the horses to stand glumly against the hedge and peer reproachfully at me through the branches when I flatly refused to bring them back to their stables, at least during the day. I was less successful at hardening my heart against a caller at the back door that night.

Not many people knock at the back door. The front one is so much handier to the yard because the original builders of the cottage didn't go in for back doors. Wise people, they knew that the back was where the draughts would come from and indeed, many such Welsh farmhouses were built with their doorless backs to the worst of the winds. There's an old story about it in a book of Welsh fairy stories I found years ago languishing in a junk shop. At least the story explains why one particular house *did* have a back door.

The owner of the house suddenly found that his cows fell sick and, in those days, before a handy rural vet was about, naturally assumed they were bewitched. As indeed they were. It was all because of his slovenly habit of tipping the tea leaves out of his front, and only door.

Unbeknown to our farmer, a family of the local Tylwyth Teg, or fairies, had built their underground home right outside his front door, so that when he gaily flung the tea leaves about, they went straight down their chimney and put the fire out. After a great deal of misunderstanding and irritation on both sides, both parties finally met up. When all had been explained, the farmer built himself a back door to tip his tea leaves out of and his cows immediately got better.

Actually I've always thought that was rather extreme. He could have saved himself an awful lot of bother and draughts if he'd just walked up the garden a bit further with his tea leaves. However, I can't really criticize because when I put a back door in this cottage, I managed to put it right in the path of the south-westerlies which can now whizz straight through and out the front door. The Griffiths, from whom I bought the farm, had put in a tiny back door already, right up at the far end of the long lean-to which was a dairy and wash-house combined. When I turned all that into a bathroom, lobby and kitchen with windows and a door dead-centre, the ghosts of the house must have despaired at my stupidity and now laugh themselves silly when I go around trying to plug up the gaps with silver cooking foil.

There is still something, however, about the place that denies the existence of a back door, so I'm unused to hearing any knocks on it. When this one came, somewhere about ten o'clock at night, I was puzzled and a bit worried. I gathered the dogs around me in a screaming pack, switched on the outside light and opened the door a crack.

Standing on the path, his back to the driving rain and gazing about him furtively, was a soldier.

'Here missus, turn that light off will you! I'll be seen,' he implored.

'Why shouldn't you be seen?' I shouted above the yelping of the dogs.

'I'm not supposed to be here.'

'That's right' I cried, eyeing him suspiciously.

He stopped twitching and turning around and looked at me indignantly.

'I'm O.K.!' he exclaimed. 'Can't you see I'm in the army?'

'Anybody,' I told him scornfully, 'can swan into an army surplus shop and buy that gear.'

Suddenly my visitor ducked down, which drove the dogs mad, so that I was too busy kicking them back to worry about the way he was fumbling in his shirt collar. He drew out some sort of disc and waved it at me.

He looked so silly squatting there in the rain that I got the giggles.

'What do you want then?' I asked. 'Are you AWOL or have you deserted or what? Don't move or the dogs will have all your moveable bits.'

The soldier darted a quick look behind him again. 'Look, the fact is, me and a couple of mates are on one of these survival exercises. We're supposed to live off the country, only there's nothing to live off and we're starving. You wouldn't have a bit of bread or something for us would you? And please, please turn out that light! They've got observers everywhere!'

My trouble is I've seen too many old movies. You know the kind, where a prisoner of war is on the run and his only hope of making it to safety is the trust and benevolence of kindly householders who will sustain him in his hour of need. So I forgot for the moment that the defence of the realm might be ruined forever if this soldier and his friends failed in their appointed task of surviving by their own wits and poaching abilities:

'Bottle of milk, some bread and cheese and apples do you?'

'Marvellous,' breathed my hero. 'For three of us?'

'All right, for three of you, on one condition. You go on your way and don't bunk down in one of my barns. I mean that, or you won't just have your observers to worry about!'

'Swear to God,' he agreed fervently. 'We'll be away and over the hills and you won't know we've been.'

'Right,' I said, 'you go back down onto the yard and I'll give you a whistle when I've got your food ready.' I shut the door firmly, if only to calm the dogs who were growling seriously now in a way that means business far more than their yelping. I should have believed them that here was a man not to be trusted.

When I gave a low whistle a few minutes later, my army friend came back, bent double, up the path and I furtively handed him a plastic bag full of cheese sandwiches, apples and biscuits and a lemonade bottle full of goat's milk.

'Now remember,' I whispered (funny how you whisper even in a gale when you're up to no good), 'you go out of that gate and steer clear of my barns. I won't have my animals upset and after all you're supposed to be roughing it in case the Russians come.'

'Honest we wouldn't dream of going into your barns,' he assured me. 'Now if anyone asks, you haven't seen us. Right?'

'Right,' I said and watched him as he wriggled away down the path clutching his supplies.

'Damn,' I said to the dogs, 'that was the last of the cheese.' They looked at me reproachfully. They're very fond of cheese.

The sheep were equally reproachful and their lambs, molehills or not, decidedly damp the next morning. My gallant band of fighting men had evicted them from the hayshed which is now their shelter and, when I went to inspect it, I found three deep hollows in the straw, the empty lemonade bottle and the plastic bag full of crumbs and apple cores. I also found the best pocket knife I've ever had, real Army issue. And if there's one thing you need on a farm and without which your life is one long frustration of not being able to cut baler twine off gates, clean out hooves, lever out old screws and do a million other jobs, it's a really good knife.

I found out some weeks later that I was not the only farm to be approached that night by various would-be survivors and not everyone was as amused as I'd been. In fact there were quite a few complaints to the army and a bit of a stink about it all. I was just glad that Mrs P hadn't been here or she'd have had them in by the fire entertaining them and we'd probably have been up for aiding and abetting fugitives or something.

As it was, I spent the next day mad with worry from the moment I knew she'd stepped onto the plane in Sydney. Of course her plane was delayed somewhere in the Middle East and I didn't believe a word of the assurances the cheerful chap manning the phones at Quantas gave me. By the time I'd finished with him, he too was beginning to doubt that the flight would ever make it to Heathrow.

As usual I was in no fit state to meet her, or brave enough to risk driving her back to Wales, so again I arranged for a car to collect her. She emerged from it onto the yard and embarrassed the driver into the ground by flinging her arms wide at me and crying 'My Baby!'. I felt a bit hot under the collar myself.

Mrs P showed no signs of jet lag and utterly refused to hop into the bed I had ready warmed up for her. She was, to put it mildly, on a high, and her adventures when the plane had been grounded with engine trouble, and her infinity of thumb-nail sketches about her fellow passengers, took hours to get out of her system. At last however, I had to go outside to feed and milk and she stretched, yawned and said, 'I think I'll just put my feet up here on the sofa till you get back.'

I came inside again about an hour later and stopped by the door, appalled. Mrs P was lying on the sofa with her face contorted and making the most awful choking noise. Before I could move, she suddenly relaxed and a look of great joy and peace spread over her face. She opened her eyes and seemed to have trouble focusing them for a moment.

'I've just had the most extraordinary experience,' she said slowly. 'I dreamt I was literally choking to death and then I felt a great release and a wonderful sense of freedom and for a moment it was as if I was running like the wind.'

I watched her thoughtfully as she sat up, threw the rug off her legs and groped for her slippers.

'Where did Merlin die exactly?' she asked quietly.

'Exactly where you were lying,' I said. 'I piled up his blankets and cushions on that end of the sofa so I could sit beside him and keep his head up.'

Mrs P nodded. 'He's alright now you know. I think he wants you to understand that.'

I didn't question what she'd said. Mrs P has always been a bit psychic. Living with her, I've got used to having my thoughts read and actually put up a kind of mind shield if those thoughts aren't too lovely. It has its uses though, because she always knows when to put the kettle on if I'm on my way home. That of course is plain telepathy, but she also has a nifty knack of being able to forecast the odd minor happening, not on a national scale, just when it affects her family and friends. Not that she tries consciously to do it; not since the time when she was very young and was fooling around reading a friend's teacup. She saw quite plainly that they would have a fatal accident a few days later and there was nothing she could do or say to stop it. It frightened the life out of my mother and she's a great one for quoting that bit about the Angel of Mercy holding down the curtain of the future. I know what she means because that's what I wish the weather forecasters would do sometimes.

That night I had the best sleep I'd had for months. Knowing that Mrs P was tucked up in her room across the landing and that now the Blight would think twice about coming back, I went off as soon as the light was out. But I too, had a weird dream.

Years before, when I'd lived in London just off Campden Hill, one of the great joys of my life had been my friend Muriel. She was married to an even older friend of mine, Jack Sassoon, who owned the amazing little secondhand bookshop at the top of Kensington Church Street. Muriel managed a small antique shop just opposite and both of them lived in a flat above that shop; a flat crammed with their varied collections of Japanese prints and porcelain cats and their real life cat, a huge black tom who always slept on their bed in a bright green, knitted pyjama suit. Jack is a highly literate Jew; apolitical, utterly irreverent, with a passion for rather naughty horror stories, especially if they feature the odd vampire. Muriel was a devout Roman Catholic with an even greater passion for racing, politics and the radio, and when she wasn't at Mass or sending telegrams to the House of Commons or ringing up her bookie, she was on the phone complaining to the BBC. I believe they had a whole file on her once at Broadcasting House.

There are many things in the cottage to remind me of Muriel; the long-handled, beautifully carved wooden bellows in the inglenook; the blue and white plate with greyhounds on it hanging on the stone wall; the pewter arms-plate with the unicorns, hooked onto the big ship's beam above the fireplace; the china ducks in a poke bonnet and a boater; a beribboned cat playing the piano and a sheep, winsome of expression, of no known breed. All of them were given or sold to me, at cost, by Muriel. I can see her still whenever I look at them; just coming up to her sixtieth birthday; clad in a shocking pink suit;

blonde hair curling over her shoulders; deep set, Plantagenet eyes in her long face like a carven saint's, sitting at the back of the antique shop where endless items of Victoriana were presided over by a big, sea-green witchball, hanging in the window, reflecting everything upside down.

Muriel would be on the phone to her bookmaker, the day's sporting page in front of her, a portrait of Lester Piggott on the wall above and the radio gossiping away in the background. She'd wave at me and then, her bet safely on, bring me up to date on her world's affairs.

'Did you hear that fool of a man this morning on the radio? Bloody politicians and the BBC should be ashamed of themselves giving him air time. I rang up straight away and complained of course! And I haven't a clue what's going to win the 2.30, so I've just put something on Lester's horse. Oh, and I want you to see this oil lamp that came in this morning, it looks just right for you! Where's Merlin?'

And woe betide me if he wasn't with me, for her other great passion was my dog and she seldom came to the Catholic Church around the corner from my flat, without coming in to see him and bring him forbidden treats: Jack was usually with her, sitting outside the church when it was fine, or in one of the back pews, wickedly reading one of his vampire books.

Muriel was a staunch Lancastrian with an accent to prove it. She was also a fervent Conservative, but that didn't stop her hiring Kensington Town Hall to hold an Anti-Common Market meeting. She paid for it out of her meagre earnings from the shop and a few lucky wins on the racecourse and she never really forgave me for not going because I was too busy meeting a deadline for the dreaded BBC.

She died, just before I left London for good, and they say that half the shopkeepers in the Church Street cried like babies. I was glad I was going away because it would have been hard to walk past that shop every day, knowing that Muriel was no longer there beating the hell out of the bookmakers, the BBC and the politicians.

The night Mrs P came back, I dreamt about Muriel for the very first time. She had on a bright pink jogging outfit and was holding a very heated argument with a sheepish looking man who bore a strong resemblance to Lord Reith. Suddenly she turned and gave that brilliant smile she always reserved for those she loved. Trotting towards her was Merlin, in a pale blue racing silk, the colour for Trap Two, which had always been lucky for him. As he joined her, Muriel nodded at her adversary and then together, she and Merlin broke into a run.

I was still laughing when I woke up, to see Mrs P standing over me with a cup of tea.

Chapter 5

The jet lag hit her two days later with a vengeance. Not that she showed any positive, outward signs of it, instead it manifested itself in a way that caused her the most humiliation. And it's not easy to humiliate Mrs P.

Before I'd known she was coming back, I'd arranged to do an interview, for his publishers, with Ian Wilson about his new book *The Turin Shroud*.

The Shroud was a subject which had fascinated me ever since the days when Mrs P had parked me in a rural convent and left me there longer than at any of the other seventeen schools I'd patronized over the years. I was there long enough anyway, to fall deeply under the influence of the Sisters of Mercy who ran the place, and to survive several 'retreats'. Before a retreat was about the only time the meagre school library was opened and we were allowed to select some religious work to sustain us through three days of total silence. For hours before the library opened, the queue stretched along the corridor, the lucky ones in front getting first pick of the gorier lives of the martyrs and the rest of us having to make do with a sliding scale of tomes on various miracles down to boring old sermons. At one retreat I'd hit the middle of the queue and managed to get the last miracle book which had a chapter about the Shroud in it. Of course there was no doubt in the author's mind that Christ had left his imprint on the shroud he was buried in, but I'd often wondered about it since then. And now I was going to have the chance to put a lot of searching questions to Ian Wilson, who had made the Shroud kept in Turin his special study, about whether it was real or a clever medieval fake.

Thanks to Mrs P's jet lag they were not as searching as I would have liked.

The Wilsons lived in Bristol and had decided that a run out to the country would do them good so we arranged for them to come to the farm for the interview. They were scheduled to arrive for afternoon tea and Mrs P bustled off to the kitchen to make some of her famous cream scones, which are usually as light as puff-balls because she makes them with cream instead of butter. She produced them with her usual modest fanfare and the Wilsons fell upon them gratefully. I was surprised that there were no enthusiastic remarks about the scones until I tasted one myself. They were like rocks and the only miracle that happened that day was that Ian Wilson and I managed to do an interview at all as our digestions went wild.

Mrs P was horrified. If there's one thing she can be proud of, it's her cooking and even the slightest failure has her in paroxysms of misery for hours, nay days, later. I've actually known her to get up in the middle of the night and make something again to find out what had gone wrong. This was the very first time she'd made a batch of dud scones however. She was incredibly silent until just before the Wilsons left. As she was saying good-bye to them she suddenly brightened and exclaimed:

'It's the JET LAG! I haven't lost my touch, it's just the jet lag catching up with me!'

Ian Wilson looked at her gravely.

'Jet Lag Scones' he said, 'that explains everything of course.' And finally Mrs P came off her high and went to bed for the next twenty-four hours to get her equilibrium back.

Someone else who was a bit off-balance however, was Blossom, the Gloucester Old Spot pig. I'd noticed for some time that she'd been remarkably slow to hear me call her for her feed and, when she came, had a pronounced list to one side. Apart from that she was her normal, cheerful, greedy, affectionate self and rolled over obligingly for me when I examined her. As she did so, her big floppy ears, which usually hung straight down over her nose, so that if you wanted to address her directly you had to pick one up and confront her eye, flattened themselves back against the ground. The trouble was immediately apparent. I have never known any animal to get such dirty ears as Blossom, but this time she'd excelled herself and they were jammed full of gunge.

'You'll have to stay deaf and wobbly until tomorrow,' I informed her straight down one of the big ears. 'I'm not rooting around in that lot tonight.'

'Eh?' she grunted.

'Forget it,' I said and gave her a slap on the rump. She rolled around, stood upright, shook her ears till they made a slapping sound like a football rattle, and picked her uncertain way back to her shelter. Soon, a mound of old bracken was shoved to the front of the shelter and a deep, contented snoring came from within.

Fortunately it was fine the next day and Blossom was already basking up on the hillside when I approached her with a bucket of warm water, a large bundle of clean rags, a nail brush, a little bottle of olive oil and a jar of a disgusting yellow ointment which not only heals small wounds, but keeps the flies off them.

There's a good old Australian expression 'In a pig's ear' and I don't understand why it should be a term of contempt and disbelief. There is absolutely nothing contemptuous about the involved convolutions of a pig's ear. Blossom had managed to get muck down every single one of hers and we spent a merry hour between us investigating those deep, mysterious channels and getting rid of what we found; Blossom in a state of utter bliss while I peered and cleaned and washed and gently oiled. Finally I adorned the backs of her ears, which already had little sun blisters on them, with the bright yellow ointment and managed to get a lot of it on me as well, which was a nuisance because it takes almost as long to wash out as the purple spray I use on the sheep's feet and which always finds its way inexorably under my finger-nails.

Blossom gave a disappointed grunt when I'd finished her ears, and then a look of sheer amazement appeared on her face as the sound of the world outside penetrated. She got up, shook her ears, winced at the noise they made, and pointed her long snout up the hill where the lambs were playing their usual game of hiding from their mothers and pretending they were lost so that the air rang with anxious maternal bleats and obviously phoney cries of infant alarm.

Blossom grunted joyously and broke into a rumbling trot up the hill, the flabby trousers of skin on her behind wobbling frantically as she went. She was, by then, getting on a bit and had lost the smooth, round rump of her youth. Watching her go, I reflected sadly that it really was too late to breed from her now. Over the years, attempts to mate her had been frustrated by a succession of problems, like no available boar within handy travelling distance, a ban on all pig movements in the area and a host of other reasons. Her unfulfilled state shocked my farming neighbours and Blossom herself agreed with them, for the only blight on her life was the lack of something to lavish her maternal instincts on.

Lambing time was a mixture of delight and sorrow for Blossom. She loved the lambs, but their mothers furiously made her keep her distance and discouraged the lambs from using her vast bulk as a jumping-off mound or their 'whippy post'. Blossom would have been more than happy to oblige and often plonked herself down invitingly near their games, yearning after the lambs, her great soul in her little piggy eyes. Now, she made her way hopefully up the hill towards the sound of their baa-ing and, as I left her, I wished her luck.

It must have been about an hour later when Mrs P said:

'There's a lamb in trouble out there.'

'No,' I said firmly, 'not in trouble, just tormenting its mother. They were all at it a while ago.'

Mrs P looked doubtful and went to open the back door. Sure enough, down the hill came the sound of genuine lamb hysterics and I swore fiercely.

'I suppose some little twerp has got itself through a hole in the fence again. There are times when I hate lambs! If they can find their way through, why can't they ever get back again?' and I marched outside, grinding my teeth.

As I reached the gate into the top field, I saw Blossom descending the hill, her bright yellow ears flapping wildly, as she shepherded in front of her a small, very frightened lamb which was screaming desperately and trying to get past her and back to safety.

Like a lot of fat people, Blossom was very fast on her feet and she headed the lamb off at every turn and succeeded, as I watched awestruck, in getting it to her shelter. She glanced briefly over her shoulder as a cry of rage floated down the hill, which was suddenly breasted by the entire flock of sheep hurtling to the rescue with the rest of the lambs leaping and hopping behind them. Blossom gave a final satisfied grunt, pushed her captive into the shelter and placed her massive bulk firmly in front of it.

The sheep continued on their rush down the hill, raging like furies, and came to a skidding halt at the shelter, and then stood darting their heads at Blossom and stamping their feet at her ineffectually. At last, the mother of the lamb, demented by the bellowing of her child inside the shelter, charged Blossom broadside on. The pig gave a deep, warning grunt and showed her fangs.

Nonplussed, the sheep stood there conferring for a moment before they began circling the shelter and Blossom like Red Indians round a wagon, while the captive within subsided into miserable hiccups. It was time to put an end to the little drama and I nipped back to the barn for a bucket of nuts.

For a moment, Blossom's greed fought with her maternal instincts, but finally she inched her way far enough from the shelter for the lamb to make its escape and rush wildly to the bosom of its frantic mother. The rest of the flock enclosed them and rapidly made their way, complaining loudly about that fat slob of an infant stealer, back to the safety of the land at the top of the hill.

I moved the sheep onto another field later that day, much to Blossom's sorrow.

'It's your own fault,' I told her, 'kidnapping's a very serious crime. But I'll put Doli and Gwylan back on this field and that will cheer you up.' For if Blossom loved the lambs, she hero-worshipped the horses. The feeling, once again, was not mutual, but at least the horses tolerated the pig (which was lucky because most horses loathe pigs) who followed them around patiently, lovingly and hopeful of their largesse. In Doli's case it was her feed nuts which she'd always had a habit of dropping untidily from her mouth when she's eating, and which Blossom had learnt to nip in and gobble up without getting kicked.

'Nonsense,' said Mrs P, when I thus explained Blossom's seeming devotion to horses. 'She sees Doli as a highly successful flirt and wants to know her secret. Poor old thing,' she sighed, 'fancy kidnapping a lamb just to have something of her own to love.'

I looked at her suspiciously and said very loudly and very firmly, 'No!'

Only someone who knows my mother as well as I do would have seen that the conversation was turning onto dangerous ground, so exquisitely subtle is she in her various campaigns to get something she wants. Currently she was putting in a bid for, of all things, a chihuahua.

It was an idea she'd brought back from Australia where she'd seen a brace of chihuahuas guarding a car against all comers. She'd had no idea before that such tiny dogs could be so fierce, and the thought that such devotion to duty could be carried wherever she went in little more than a large handbag had sparked off Mrs P's ambition to own one. It was so unlike her actually to want an animal of her own, that I almost gave in, but there was the dog itself to consider.

'A chihuahua would die of cold here in this cottage,' I'd told her. 'They need a lot of warmth. Besides which, you're already surrounded by dogs who adore you.'

'They're *your* dogs,' she muttered defiantly, 'I just want one of my own which wouldn't take up much space or want too much exercise.'

I hooted. 'You have to be joking! The last pair I saw never stopped exercising . . . all over the backs of the furniture!'

The pair had belonged to the actress Irene Handl and they had kept up a dizzy, perpetual motion while she and I had been trying to do an interview about, fortunately, the Stars and their Dogs. Miss Handl had been the one to warn me about the need to keep chihuahuas warm and had sent me away with a photograph of her two. Mrs P, a great fan of the actress, had commandeered it and that probably began the first stirring of this sudden obsession of hers.

If I could, and she wanted it, I'd give my mother the moon, but there was no point in giving her a chihuahua if it was going to die of cold before she'd had it five minutes, so I sternly headed off any return to the subject. Apart from the odd moments when she was moved by something like Blossom's desire to have her own small companion, Mrs P went very quiet about the Chihuahua Project and diverted her attention to other small creatures, like Mima's ducklings.

Mima was sitting cosily on twelve eggs in the big hen ark at the back of the house. She was an arrogant muscovy duck and, although she was an utter pain in the neck in many ways, she was the darling of my eye because she was my very first success story with an animal.

Mima went right back to those days when the Artist and I were bumbling about trying to decide what to do with the farm. Apart from a few chickens, which someone had donated to us, we possessed no stock and, with the mistaken idea that they'd be good for keeping the grass down, we decided to acquire a pair of geese. I have told the story elsewhere of how we got those geese, Douglas and Daisy, but just as we were leaving the farm with both of them subdued in sacks on the back seat of the car, a self-important hen came strutting past with five tiny ducklings drifting after her.

Cliff Griffiths, who had sold us the geese and indeed my new farm itself, waved a hand at the hen.

'She had six this morning,' he declared. 'She must have lost one somewhere.'

I paused before getting into the car. 'How do you mean, lost one?'

'Well, I don't think she lost it, just left it behind. I noticed one of them didn't look too clever this morning.'

'Did you do anything for it?' I demanded.

'No point,' said Cliff, shrugging, 'if it's no good when it's only a couple of days old, it's a waste of time.'

I don't know what I thought I was in those days, Florence Nightingale of the Animals or someone, but I insisted on Cliff showing us where the sick duckling might be.

He took us round to the back of the barn and there, sure enough, lying on its side in the nesting box, was a scrap of black and yellow fluff. I picked it up sadly and then, 'It's still breathing!' I exclaimed.

Cliff grinned at me sarcastically. 'Want to take it home then? Save it and everything? It'll never be any good! The hen knew best to leave it.'

I gazed at him sternly and appealed to the Artist, who was

completely of my opinion that we should take the pathetic bundle home and let it die in the warm instead of alone in that cold box. It never occurred to me that it would live, and I dribbled brandy and water into its bill just to feel I'd done something to help it. I kept dribbling it in all night and by the next morning the duckling was standing up imperiously in its box demanding proper food and every attention.

I gloated over Cliff for years because of Mima, and he, for once, conceded that I knew a thing or two when he saw the big, sleek, black and white duck waddling up and down my yard. Unfortunately there were no little Mima's to follow because, after a strenuous fortnight with my foundling, the drake we bought for her, keeled over and died. When yet a second did the same, I decided Mima was too much woman for them and gave up.

Muscovies can be very bloody-minded ducks. They are far wilder in their nature than the ordinary domestic duck and still retain their ability to fly high and wide. Mima, balked of her legitimate mates, refused to have anything to do with mere Aylesbury crosses or the Khaki Campbells and took up flying seriously as a hobby. For most of the time I loved to see her swooping rather ponderously about the valley, or sitting grandly on the ridge tiles of the barn before deigning to join the rest of the flock as they came in for the night. In spring, however, Mima drove me mad by electing to sit out on the pond at night. If I tried to drive her in, she simply took wing until I gave up. I usually won in the end, but it was often midnight before she capitulated, when it was too dark for her to see where she was flying and cold enough for me to curse her as I blundered into the pond and got water down my wellies; until the night she wasn't there and I spent the rest of it stumbling around the fields with the torch, convinced that the fox had got her, or would if I didn't find her.

She appeared, bright as a button, on the yard next morning and then, when my back was turned for a moment, disappeared again. It took me three days of crafty tracking to find her and the nest she'd made, full of infertile eggs.

Given her determined nature, I knew that nothing but a proper nest with proper ducklings to follow would end this war of wills between Mima and me, so I put her into the hen ark with some fertile Khaki Campbell eggs and gave her well-being into Mrs P's hands. At least it would give them both something to think about and hopefully give me some peace.

By the time Mima's ducklings had hatched out, Mrs P seemed to have forgotten all about chihuahuas. She was far too busy arguing with Mima about whether the ducklings were old enough to go down onto the pond.

'Those buzzards might get them,' she complained to me as Mima sulked in her pen and the ducklings skittered about trying to catch flies.

'Crows more likely,' I said, 'but the buzzards might have fancied that chihuahua if you'd had one.' I'm a great one for ramming points home.

Mrs P looked at me in shock.

'I didn't think of that!' she said. 'I suppose they might have thought it was a rabbit! And those ducklings,' she informed Mima, 'aren't going out till I think it's safe.'

Mima stuck her head out at her and hissed. She'd met her match in Mrs P.

June came in glory that year; the creamy swathes of hawthorn were thicker and richer than ever in the network of hedges climbing the hills, and the perfume of it scented the whole valley. By the time the hawthorn had turned that faintest of pinks which means its time is nearly over, and every breeze brought down showers of petals like drifting confetti, I began to be a little apprehensive.

My birthday is on the 4th July and, for some reason, is always blessed with personal disasters. It's impossible to forget it because, being American Independence Day, the press and radio remind me of it, so I just grit my teeth and endure. I try to stay put that day, to keep the swathe of misery enclosed, and sensible people stay away from me. That year, however, I had a funny idea that if I treated it as a day for giving rather than receiving, I might just lift the jinx. Perhaps if I gave Mrs P a present (for after all she'd done all the hard work originally) it would take my mind off my birthday. The trouble was, what to give her. I was tired of buying her expensive things only to have them languish in a drawer for ever, and the only thing she'd hankered after recently, was that chihuahua.

One evening, a few days before the dreaded 4th, Mrs P retired to the kitchen and shut the door firmly behind her. I knew she was making a birthday cake in spite of my command to ignore the whole rotten business.

It was Friday and, for some reason, I'd picked up a copy of *Horse and Hound* when I'd been to town to do the shopping. Thumbing through the classifieds at the back, I glanced quickly down the 'Dogs for Sale' column. There, amongst the pointers and labradors and border terriers, leaping at me from the page, I read:

'Chihuahuas. Three smooth coated puppies.' There was a phone number and it wasn't a million miles from here.

There was a whiff of sulphur as Mrs P's fairy godmother zipped past and I thought, 'Well, it wouldn't do any harm to just *ring* them.'

Chapter 6

'WHAT sort of chickens did you say?' cried Mrs P.

'Er . . . White-Crested Polands,' I said tentatively, waiting for the worst of the storm to subside. 'Gwynneth and her kids and I are going over to Hay-on-Wye to collect them tomorrow.'

Mrs P glared at me in disbelief.

'I don't know what to make of you,' she said furiously. 'You were only complaining the other day that you were sick of chickens rooting the place up. And wasn't it that horrible Poland rooster you were chasing around like a mad thing?'

She was right actually. Polands are beautiful to look at, the hens with incredible big, soft pom-poms encasing their heads and the cockerels sporting an amazing confection of feathers that would make an Indian chief jealous. They don't really come from Poland. They get their name from the extraordinary bulbous bone structure which is part of their skull, unlike other poultry breeds which have crests that are merely an accumulation of feathers on a fleshy base. Apart from this, the Polands come in the most lovely colours from blue and white and black to an exquisite powdery-buff and, like the ones I had, a rich gold on which each feather is outlined in black. They're completely mad birds, although, to do them justice, that's because most of the time they can't see where they're going and if you clip their crests back a bit, they are more sane. They're a breed, however, which needs a lot of attention, because those spectacular crests get wet and soggy or become infected with mites, and really they're a specialists' bird. Of course no one told me that when I bought mine, captivated at the time by their glamorous looks rather than giving any thought to how I'd look after them.

I'd lived to regret my impulse, not so much because of the little hens who meandered about short-sightedly, but because the rooster and I had a screaming match nearly every night. I dashed around the yard after him while he shrieked and leapt as if he was being murdered, and flung himself further up the field, far from the safety of the poultry shed. In the day, we went through the same routine in reverse, with me trying to evict him from the goat-house where he was merrily chucking their bedding all over the place. If I locked him up in a pen with his wives, in an effort to get them to breed true, he sulked in a corner, ignored them and refused to eat. One way and another, we didn't like each other very much, so Mrs P's scorn when I mentioned that I was off to buy yet another trio of Polands, was understandable.

I realized, too late, that I should have chosen another breed to explain why I was going off with the large wicker cat-basket on a totally unscheduled trip with Gwynneth.

The night before, Mrs P, who has ears like a lynx when she wants to, had heard me muttering and whispering into the phone, even though she was supposed to be locked up in the kitchen with her cake-making, and had drifted out casually to see who I was speaking to. I'd warned the man on the other end of the phone that this might happen.

'If I suddenly start talking about chickens,' I'd said, 'just go on as if I was still discussing the dog. I don't want my mother to know I'm ringing up about a chihuahua.'

So when, as I'd expected, she did emerge to cock an ear, I hastily said: 'Well now I presume these White-Crested Polands are in good health?'

I was referring to that most dramatic of the colour combinations in Polands, in which the body is blue or black and the crest is white. I'd been telling Mrs P about it just a few days before, and so it was the first type of chicken I could think of.

'In perfect health,' said the voice on the end of the phone without hesitation. 'As I was telling you, we bring them up tough because this is a farm, so my wife whelps the bitch down in an outside shed and that's where the pups are brought up. Oh yes, they're real farm-bred chihuahuas!'

I glanced up and saw that Mrs P had retreated once more to the kitchen in disgust and shut the door loudly behind her.

'What colour are they?' I asked, lowering my voice again.

'Well two of them are black and tan like the bitch, but there's one quiet little chap which is a bright tan colour with a funny half-collar of white.'

'Can I come and see them tomorrow?'

'Yes of course . . . Now, you take the road from Brecon to Hay-on-Wye . . .,' and he went on to give me precise instructions how to find their farm.

I put down the phone and then realized that I'd got carried away too soon. My old blue Beetle was, as so often those days, languishing on the yard with something wrong with it. I picked up the phone again and rang Gwynneth Griffiths who lived across the valley, was a great friend of Mrs P's and always game for something new. When I told her the story and my problem, she immediately offered to drive me over to Hay-on-Wye to see the pups.

'We're going to get chickens, mind,' I warned her.

'Right,' she laughed, 'I'll tell the kids. I think Siân and Dafydd would like a run out too, being as it's Saturday. They've never seen a chihuahua!'

Mrs P's scathing remarks about chickens, crested or otherwise, as we set off into a bright golden day next morning, rolled off our backs and made Siân and Dafydd giggle uncontrollably half way to Brecon.

'Won't she be surprised when we come back with a DOG!' Dafydd chortled, hugging himself with glee.

61

Siân was more scientific about it all. 'How big is a Chic-Choo pup, Jeanine?' she asked.

'I don't know,' I told her. 'Very, very tiny I expect. Like a minute rat I suppose.' And for the rest of the journey, they speculated and discussed the size of the pup until it had almost diminished from possibility by the time we'd arrived.

They were in for a big disappointment. The pups were the size of very large, overweight guinea-pigs and the parents more like tough little terriers than the airy fairy chihuahuas I'd seen before.

Siân and Dafydd looked at me reproachfully.

'It must be because these are proper farm-bred chihuahuas,' I improvised. 'The very tiny ones are city dogs!'

Being sensible, farm-bred children themselves, they accepted this, and by now the antics of the two larger pups had them fascinated. They rolled about and chased each other and never stopped moving and showing off. Sitting on the sidelines, his head poking out of a blanket, the little tan pup with the half-moon of white around his neck, looked at them with contempt. When the three of them had been brought in from the shed with a blanket round them, the two black and tan ones had promptly tumbled out of it and begun to entertain their public, but the other one had quietly commandeered the blanket, dragged it to the side of the room and proceeded to bury himself in it. Instead of joining in the general admiration of the two little exhibitionists, I crept over and pulled the blanket right back from the tiny face with its big, melting eyes. The lips curled back and, as I put my hand towards it, the pup bit me. It hurt too.

'Oh you don't want that one,' said the owner of the pups. 'He's a funny little thing, just keeps to himself all the time and goes off on his own with that blanket.'

Now the golden rule, they say, for buying a pup is never, under any circumstances, to take the quiet, retiring one that stays away from the pack. It is more likely to be unhealthy. But there was nothing unhealthy about the gleaming coat of the little tan pup. If anything, he looked fitter than his brothers. What he did remind me of, was a photograph I have of myself when young with a group of other five-year-olds, all laughing and dancing about in front of the camera. I'm standing there glowering in the middle and I remember that day very well, thinking what idiots they were and determined not to play to the gallery. The pup had pretty much the same expression on his face. Suffering fools gladly was not his style. He and Mrs P would get on very well.

When we arrived home that evening, I carried the wicker cat-basket carefully in front of me, while Siân and Dafydd walked behind making loud clucking noises.

Mrs P emerged from the kitchen. 'What have you brought those chickens in here for?' she demanded.

'Thought you might like to see them before I put them in the barn,' I said and placed the basket carefully on the sofa. Siân and Dafydd increased their chicken noises and hid behind the door.

Mrs P groaned. 'All right then. I'll have a look, if you insist, and then you take them outside and have your dinner.'

There was a muffled giggle from behind the door and, as Mrs P bent down to peer into the wire front of the cat-basket, the two children emerged with grins all over their faces, desperately waiting for Mrs P's cries of surprise and excitement.

They waited a long time, for Mrs P was totally dumb. She gazed and she gazed and then she reached into the basket for the pup and from that day on, to think of one without the other was quite impossible.

She called him Winston and it fitted him perfectly; small and round and immensely conscious of his own dignity as he is. The whippets, after an initial hope that he might indeed be a rat and legitimate prey, were rapidly disillusioned and, in fact, by the time he was a year old, Winston was undisputed leader of the pack. He is the only dog I've known who would stand up to the goats and even his first encounter with the immensity of Doli fazed him not one bit. Anyone who has ever watched Winston careering around the fields after the whippets, or tunnelling through a snow drift, or rolling gleefully in a mound of sheep muck, immediately revises their opinions of chihuahuas as delicate little toys. Any more doubts are quickly scotched by a quick nip from Winston if he thinks they're laughing at him. He is, of course, a solid ten pounder, not one of the little show specimens that weigh in at about four or five pounds, and Mrs P had to give up the idea of carrying him around in her handbag. Apart from the weight, the indignity of it would have had Winston in a fit.

His dignity is everything to Winston. Myrddin calls him the PM and when Winston potters about with one of those little cylindrical dog-chews stuck out of the side of his face, you can see why. His self-appointed task in life is to keep the whippets in order and we have to be very careful if one of them is being told off, for Winston moves like a ferret and jumps like a bean and, immediately ranging himself on the side of authority, he'll bite the offender's behind and hang on. He does, however, let us know if one of them is in trouble, or needing to go out, or if their water bowl is empty. His other useful trait is to answer the telephone, or at least let us know when it's ringing.

He developed this trick to help Mrs P. Often I would hear the phone ringing when I was outside, steam in, expecting Mrs P to be answering it, only to find it gone dead and my mother out in the kitchen with the door closed and the radio on full blast. Panting and cursing, I'd storm about and tell her to leave the wretched door open or turn the radio down so that she could hear the phone. It might be work or something.

Winston, who has ears almost as big as himself, was Mrs P's comfort in these rages of mine and she would complain bitterly to him about me. He must have taken it in too, because he solved her problem by barking at her furiously every time he heard the phone, running round her feet, rushing to the door and urging her frantically to get moving. Once the door was opened, he'd rush to the chair by the phone, hop up, put his head up against the instrument and howl to let it know someone was coming.

His worst fault was, and still is, rolling in every disgusting thing he can find and, when you're as small as Winston, that means having a complete bath in it. Which is why he also answers to 'Pongo'.

Of course, when he first arrived, he didn't get much chance to roll or even have a good wallow in the grass. Mrs P's conviction that the buzzards would have him was obsessive and she could have been right, for they wheeled overhead endlessly, searching the fields for food to take home to their newly hatched offspring in the big oak down in the wood. My birthday had passed off without too much trauma, apart from a spectacular combination of wind, rain and fog but, as the jinx sometimes lasts for a few weeks, we both decided that one really massive awfulness would be for the buzzard to collect £40 worth of chihuahua and give it to its children for breakfast. The worst thing that happened, however, was that Doli's daughter, Gwylan, went missing.

Doli herself was away, hopefully being schooled by a lady who did a bit of show jumping and took a few riding pupils.

It was quite a long time now since Doli, the big draft mare, had wormed her way into my sympathy and been bought to save her going back into the forestry to work. She had bred three foals, of whom Gwylan was the latest and was now a strapping yearling completely subservient to her mother.

Doli had originally been using up some spare grazing on my land when the trekking centre which had been using her to carry heavyweight trekkers about had found her less than reliable with her riders. In those days, fairly fresh from a hard and none too kindly life, tushing timber in the forestry, Doli had been an amiable enough soul, but years of being spoilt hopelessly by me had accentuated her rather uncertain temperament. Certainly I didn't feel up to riding her again, especially when one experienced rider had taken her out for a trot round the hills and come back with the pronouncement:

'It was like riding a big coiled spring and I'm only thankful it didn't go "Boing". I'd have her schooled a bit if I were you before you ride her again.'

And so to be schooled she'd gone, while I concentrated on getting Gwylan to lead properly without her mother inciting her to rebellion every time I tried to walk her up the yard.

Gwylan went looking for her mother two days later. She'd have gone looking a lot farther than she did if she hadn't had trouble negotiating a bit of wonky fence, backed up by an overgrown hedge, on one of Cliff's fields and been spied by Cliff himself so that we could catch her and bring her back. But it's a nasty moment to look out onto a field that has recently been fully occupied by a large horse and a few minutes later see it completely empty. Gwylan, however, was not of a persistent nature, just a bit lonely and, remembering that Blossom was always good for a laugh and a bit of teasing, finally made do with her, even standing over the pig shelter at night and nodding off in time to Blossom's snores. The reports about Doli were likewise reassuring.

'I don't know what you've been fussing about,' said the riding school lady. 'I've had all my pupils pottering about on her. She's the perfect beginner's horse, as long as you can get your legs round her.'

'Do you think that's wise?' I asked. 'I know Doli seems docile enough, but she does have her moments you know and she's thrown a few people in her time.'

'Nonsense! They probably didn't know how to treat her. She's an absolute angel. Everybody loves her!'

I sighed and hoped she knew what she was doing. Perhaps it was just here at home that Doli was a right old madam, knowing I'd let her get away with it. Sure enough, the reports continued to be of a willing, obliging horse who stood patiently while beginners crawled all over her and who plodded quietly and obediently along. I began to have hopes that one day Doli and I would go for long, contemplative rides over the mountain.

She'd been away about three weeks when the phone rang and someone squawked indignantly at me when I answered it.

'Your horse is coming home tomorrow!' I was told. 'She nearly killed me today. Thank God it was me and not one of the pupils! I was riding along quite calmly when she just exploded under me! Boom! Up in the air on all four feet! And she went on and on doing it. She was absolutely determined to get me off! And the mood she was in, I wouldn't have given much for my chances on the ground.'

'I gather,' I added warily, 'that she didn't get you off.'

'Did she hell! I wasn't going to be got off by a bloody draft horse! But I see what your friend meant about a coiled spring. When Doli goes "Boing", she goes "Boing". What is worrying though, is that there was nothing to make her do it. It was simply as if something just clicked in her brain and she was off!'

I knew what had clicked alright. Doli had suddenly had enough. It was all very well people drooling over her and thinking she was an angel, but there'd been a bit too much of this riding business. Somewhere over the hills was a nice mug who expected nothing of her except to eat her head off, toddle along to the stallion occasionally and have the odd foal she could boss about. It was time she went home!

Chapter 7

The fog drifted down almost casually from the top of the Fans. The great wedges of sandstone, towering over the valley, had draped it around themselves in swathes before releasing it to come lower and lower till it swirled and settled in thick, malevolent silence around the buildings on the yard. Dim shapes of horses and sheep passed beyond the field gates, the goats retreated into their shed and sat cudding in that glazed, mindless way they have. Chickens and ducks were unusually quiet, except for a loud squawk when Mima crash-landed on the pond on top of the Khaki Campbell drake and then beat the daylights out of him for getting in the way. Down in the pine trees by the river an owl, confused by the sudden gloom, woke and hooted tentatively. I hooted back, for something to do, and then cursed.

Why, of all days, did it have to be this cotton-wool world instead of the bright, glowing, summer valley of yesterday, layered in thick shades of green and blue as the woods thickened and expanded and, in the fields set aside for hay, the grasses bending and moving, silver and aquamarine, with every slight breeze? If it had been like that today, Michael Bowen might have come, but not, I thought, with the fog blearing over everything.

I shrugged hopelessly and wondered yet again why, every time there was a slight chance of getting back into broadcasting, the job I loved best, everything conspired against it. For Michael Bowen was the Network Radio Editor in Bristol and he'd told me that if the weather was good, he might just bring his family out for a week-end drive and have a word with me about one or two of my ideas.

I'd originally sent those ideas to my old producers and editors in London, where they'd met with a total lack of interest. Apart from the continual blow to my pride, it was pretty worrying because although I'd been saved from having to sell the farm, the year before, by my father's legacy, most of that had already been eaten up in settling a massive overdraft. There wasn't even enough to pay off the mortgage on the farm, unless I could sell a small flat which had been part of the legacy, but was occupied by tenants who had no intention of moving. What had seemed wonderful on paper, was rather less in reality. At the time, however, it had saved the day and, because I believed that now my luck was on the turn, I had hoped that at last the BBC would give me the programme I'd spent years planning for and I'd earn my living again.

One of the excuses I'd made, when the Artist had left the farm and I'd decided to come and look after it myself instead of working in London to support it, was that if I really learnt more about the basics of life on a small farm, I might have more success with an idea I'd been flogging around for ages. This was to do a weekly programme on such a life. Three years had passed since then and still no one was interested in that programme, in spite of the great army of people who had since left the city to take up small scale country life, either on a few acres or in various rural industries and crafts. Finally I'd given up and begun instead, to put forward 'one-off' feature ideas.

But I'd been away too long. One of the tricks of staying in broadcasting work is to be on the spot, to have your face about the place and to keep up with the politics of who is doing what and where. I was too far away and hopelessly out of date. The pages of ideas I kept sending to London came back regularly with polite regrets and hopes for my continued good health, until, in sheer desperation, someone had seen fit to send the lot on to Michael Bowen in Bristol. As far as I was concerned, that was the kiss of death. If they were passing the buck now, then London was well and truly finished with me. I didn't know anyone on the radio side in Bristol and didn't think they'd be over-enchanted with cast-off ideas from London anyway. I didn't even bother to ring Michael Bowen.

And then Michael Bowen rang me. He would like to talk to me about one or two of my ideas and it so happened that he and his family had a particular love of my part of Wales. If it was reasonable weather, they might just take a run out to see me this coming week-end. Would I be at home? Oh yes, I'd be at home! With bells on and the farm polished up to the nines and Mrs P earnestly getting on to the Hot Line to demand Good Weather! Instead, this horrible blanketing fog had descended so that everything dripped and gloomed about and the beauty spot the Bowens wanted to see could have been a million miles away in that thick, white muck.

'They won't come,' I told Mrs P. 'Him Up There has obviously not been listening to you. I mean, fog like this in the middle of summer!'

'We *do* get fog here regularly, summer and winter,' she reminded me sourly. 'Ah well, if you don't need me any more, I'll go and attend to that cottage cheese.' And off she trotted to her lair in the kitchen where a big muslin bag, full of curds, hung from a hook and dripped whey into a bowl underneath it.

Now that Dolores had kidded again, we were swimming in milk and Mrs P had been furiously using it up by making cheese. She had also made it out of the whey.

'I can't bear waste,' she'd complained when the cheese was made and the whey had nothing better to do with itself than be given to the dogs.

'The dogs don't think it's waste,' I cried. 'They think it's lovely and it's very good for them.'

'I still think there's something else I could do with it,' she mused, and buried herself in the various books on cheese-making I'd acquired in a first flush of enthusiasm.

The result of her researches was a Norwegian recipe for simmering the whey until it reduces into a thick, caramel-coloured goo. When it's set into a mould, it goes a bit like fudge. It was quite nice for the first few times but Mrs P never does anything by halves and soon masses of the stuff was languishing, unloved and unwanted, in the fridge.

'Why,' I'd finally said to her, 'don't you save the whey for me to give to Blossom? She's looking a bit run-down lately.' And that did the trick because, in her heart of hearts, Mrs P was likewise sick of making the new cheese and had been longing for an excuse to give up her saving ways.

When I joined her in the kitchen now, she was busily decanting the finished cottage cheese into a bowl and adding chopped chives to it. It was about the last time I saw her using anything so mundane as chives for a flavouring, for we were within days of the great Herb Revolution. The whey she carefully put aside in Blossom's 'specials' bucket, and I sent up a silent prayer of thanks.

We were just about to sample the fresh cottage cheese for lunch, when I heard the sound of a car and the dogs began a frantic yelling, headed by the baby shrieks of Winston as he exploded out of his little basket by the Rayburn and led the rush to the door.

There, beaming at us out of the fog, clad in hiking boots and anoraks, stood Michael Bowen with his daughter and son-in-law.

'I wasn't expecting you,' I cried, 'not in this fog.'

'Actually, it's not too bad,' said Michael, 'and I think it's clearing anyway. We'd love to have a look around, by the way.'

And for the next half hour we tramped through the yard inspecting the animals and buildings and discussing Michael's own plan to buy a farm in Wales, which his son-in-law would run. The Bowens already had a smallholding in Somerset and, to my amazement, Michael actually knew what he was talking about when it came to the joys and miseries of running such a place.

'Lunch?' I asked a while later.

'We've got our own sandwiches thanks. We know what it's like having visitors drop in when you're busy,' said his daughter, and I realized that they really *did* know and didn't, like so many others, expect that any rural establishment is automatically overflowing with vast quantities of home-made wonders.

While they munched their sandwiches (which mortified Mrs P

who was longing to run up one of her instant miracle feasts for them) and drank tea, we talked about the Bowen's smallholding, their new farm project, and from there I began somehow to declaim about my old idea for a programme on small-scale country life. By now I'd forgotten totally that Michael Bowen was a man of some note at the BBC. He was just another smallholder gossiping over the problems.

Finally they got up to go and Mrs P, smiling sweetly, tried to press a large pot of the dreaded whey cheese on them. I could have brained her. And then Michael turned and said thoughtfully:

'How would you like to make a pilot programme for your country idea? I've got a small fund for that sort of thing and I wouldn't mind having a go at getting it accepted.' I stood there, silent with disbelief and the first, faint glimmerings of hope, as they collected their things and went out of the door.

I can't remember another thing about that afternoon except dancing around and frightening the goats half to death by singing all kinds of dreadful triumphal songs to them, loudly and very off-key. But Blossom thought heaven had happened when I gave her pots and pots of caramel cheese as well as the fresh whey itself.

It may have been the return of the fog that night, this time deep and solid so that I had to grope my way back to the front door when I'd seen the stock in, and the Poland rooster for once gave up his nocturnal cavortings and made for home with only a token shriek, or it may have just been my natural pessimism, but my euphoria at this first interest in my benighted programme had quite disappeared by the time we sat down to dinner. We were having it on trays in front of the fireplace although, in spite of the fog, it was far too warm to have a fire in it.

'Even if I do a pilot for Michael,' I said gloomily to Mrs P, 'first he has to like it and then he has to convince the Controller of Radio 4 and I've been through that routine before without any luck.'

Mrs P put down her fork and glared at me.

'I've told you before and I'll tell you again,' she said, 'your trouble is, you have no faith!'

There was a loud bang and a cascade of soot came thundering down the chimney and landed at my feet. It always amazes me that I can sweep that chimney, even attack it with hammers and long-handled hoes to try and displace the soot and tar, with very little success but, when you least expect it, quite by itself it will chuck the stuff down and scatter it all over the floor.

'You see,' said Mrs P. 'Bw agrees with me.'

I closed my eyes wearily and said, 'Bw doesn't know anything about it and neither do you,' and then flinched as another clump of tar shattered itself all over my feet.

'Now you've annoyed him,' said my mother sternly.

'Mother dear,' I said patiently, 'this is serious. I don't mind your game about Bw, but when you start invoking him as an oracle, that is going too far!'

'Bw' is short for 'Bwbach'. If you consult that excellent book, *The Personnel of Fairyland* by the folklorist, Dr Katharine Briggs, you will see by the entry for 'Bwbachod': 'The Welsh brownie people. They are friendly and industrious, but they dislike dissenters and teetotallers.'

It was Katharine Briggs herself who introduced me to these broadminded sprites. Many, many years before, I'd found several of her books, on the homelier side of British folklore, on the shelves of Jack Sassoon's bookshop in Kensington. I immediately began to collect as many as I could, because they were not only full of definitions and stories about everything from the genial house hobs and the romantic trooping fairies to the nastier creatures like the Shriker or Peg Powler, but they were beautifully and wittily written. Later still, I found her children's book *Hobberdy Dick*, in which she had used her great knowledge of folklore and her love of the Cotswolds and the seventeenth century to create a small masterpiece. After a lot of nagging hither and yon, I'd managed to interest one of my producers in Dr Briggs, and arranged to go and see her.

By then she was in her seventies, living in a magical house in the Cotswold town of Burford, with her books and her Siamese cats and great landscape paintings by her father, the water-colourist Ernest Briggs, covering the walls. She was just as intriguing as her books, with her wonderful, dramatic voice which she could use to tell a story in such a way that, on that winter's day, sitting by the fire and only pausing to throw a log on it occasionally, I was pixie-led back into another time and another dimension. I was still there when her friend and secretary, Josephine, drove me to the station at Charlbury to catch the London train, accompanied by a small terrier called Daniel.

It was only a few weeks before Christmas and the little town of Burford was already glittering with decorations and that excitement which seems to spill out of shops and on to the hurrying people at that time of year. It all seemed part of the bewitching world we'd left behind at the Old Barn House. By the time we got to the station at Charlbury, there was light snow falling. Josephine and Daniel and I hurried into the waiting room where a small, coal fire was bravely doing its best for a group of frozen business gents returning from a conference.

At last, with a warning toot, the train came in, drew up with a hiss of brakes and decanted a number of tired-looking people and a large tough dog, pulling at his lead, held by a portly man in a tweed hat. Daniel, who till then had been the soul of politeness, made up his mind instantly about that dog. Just as I was about to scramble up into my carriage, he flung himself to the attack.

Josephine stood there appalled, the stout man let go of the lead and bellowed 'Heel Sir!' ineffectually at his dog, and the business gents tried to pretend it wasn't happening. Above the racket, I heard the doors of the train being slammed shut and a guard stood, fascinated by what was now an almighty dog fight, poised to blow his whistle. Slowly he lifted it to his lips.

'Don't blow that whistle! Hold the train!' I shouted at him, rushed back along the platform and pulled the warring dogs apart.

A loud cheer went up, Josephine hauled Daniel back to her car, the stout man cuffed his dog around the ear and the guard yanked me back on to the train. It wasn't till the station was slipping past, its dim lights blearing through the snow, that I realized my sheepskin gloves had a nasty tear in them and blood was oozing gently out.

I have told that story because, you see, I am the greatest coward on two legs, especially when it comes to dog fights and it illustrates the kind of spell that Katharine Briggs cast. She was the sort of person who could make you feel capable of anything and yet, at the same time, get you to laugh at yourself and the world about you and take it all with a pinch of salt. When I was with her, I'd told her about some of the odd things that had been happening at the farm, which I hadn't owned for very long then. She'd been highly amused when I told her about the odd plates and the china candlestick which had launched off into space unaided and crashed on the floor.

'Have you,' she twinkled at me, 'been paying your due respects to the Bwbach? That's the Welsh version of my Hobberdy Dick you understand. And you know they like to think someone appreciates them.'

'Are you serious?' I asked.

'Not in the slightest,' she laughed, 'but it wouldn't hurt to hedge your bets would it?'

So I'd told the Artist and Mrs P about the Bwbach and for some reason we'd located his probable headquarters around the big, open fireplace and even called out to him up the chimney sometimes. It was a comfort to think that he was there disapproving of dissenters and teetotallers, even though he didn't show too much sign of industriousness, except for this odd business of cleaning the chimney at awkward moments.

Now I sat there looking ruefully at the mess on the floor while Mrs P grinned triumphantly. And then she said rather more seriously:

'I mean it you know, about you needing to have more faith, in yourself, if nothing else. I've got a feeling things are going to get really good from now on. My God! What's that?' And there, fluttering down the chimney was a small, dark shape.

It drifted across the bare hearth and out into the room and came to rest, wings outspread, on the sofa, where the two cats lay curled up asleep together. So silent had the small creature been, that they didn't so much as twitch a whisker as it settled beside them.

It was one of the army of bats which live in the loft, hanging in bunches from the end-beam in summer and hibernating under the thick plaster, which lines the slates, in winter. I loved to watch them emerging from their launching pad on the roof on warm nights and to see them swooping over the yard and away into the dusk. Often, in the early hours, I was awoken by them coming back with tiny muffled noises just above my head as they settled onto their roosting places again. Afraid of frightening them away, I never went near them unless it was vital to go up into the dark, cramped loft, but I'd always longed to see one up close. I could only assume that this one had been put off course by the fog.

Mrs P does not share my love of bats. To her they are only one degree better than rats or mice, which she hates. Now, seeing my delight as I picked the soft, velvety shape off the sofa and out of range of the cats, she snorted.

'I hope they're not going to make a habit of this!'

As I carefully carried the bat upstairs to put it into the loft, I heard a deafening crash which stopped me dead. And then came Mrs P's furious voice:

'Listen here, you, Bw! Do that once more and by the time I've finished with you even the Fairy Queen herself won't recognise you!'

It must have been about one o'clock when the owls woke me, calling in strange sobbing moans. There was one in the big ash tree outside and, as it cried out, another owl answered it from across the valley and yet another picked up the cry from somewhere over in the forestry. The sounds rose and fell and then travelled away as one owl took flight and called as it went.

I got up and leant out of the window, high up in the pine-end of the cottage, and saw that the fog had gone completely. The sky was clear and fading stars scattered through the branches of the ash tree while an adolescent moon made its way home. The only clouds were a few stray wisps that had got caught half-way up the sides of the Fans, which now rose sharp and precise on the horizon, with a faint halo of light behind them. The owl in the ash tree stirred and silently lumbered off its branch to float down towards the river. Another faint movement below caught my eye and there, dithering silently across the yard, its timing perfect, was a bat.

═══ *Chapter 8* ═══

'Not really unique, but interesting,' said the tall man beside me.
'How interesting?' I asked him.

'Well, you've got about 53 plants here that are not so common in
this part of the world any more and could be quite endangered one
day if we don't watch out. Anyway, it's an interesting set of habitats
you have here. You've got the river bank, the unploughed pasture
land and a few, very pleasing damp patches. Then of course there's
the woodland up on the slope there. And for the sake of the wildlife,
I earnestly beg you not to uproot those big bramble patches, or
remove all those rotting trunks of wood there, my colleague who
specializes in fungi would be most interested in those. And, as you
can see, you have a nice lot of butterflies here. Again, nothing really
special but it's good to see so many of them.'

I was disappointed. I'd hoped that my meadow would yield some
really rare plants or small animals. Now that I'm older and wiser I'm

quite glad there wasn't anything to get the man from the Plant Conservation place too excited, or by now there might have been armies of them slapping preservation orders about and generally treating the place as their own and me as the humble caretaker. But in those days I was innocently wanting to do my bit for conservation and also, if the truth be known, looking for a good excuse to explain why I didn't 'improve' the meadow into the rich grazing land I saw all around me. For I liked my meadow as it was, boggy patches, brambles and all.

Now, in high summer, everything was afroth with meadow-sweet and my plant boffin had also shown me the delicate nodding heads of quaking grass, and pulled aside the reeds and rushes in a particularly squelchy place to point out wild angelica and fleabane and a lot of other plants I had to go and look up later.

You reach the meadow by slipping and sliding down a perilous slope through the small wood which begins as hazel and goat-willow and ash, and progresses to big, stately pine trees bordering the river. From there the path becomes a nightmare of displaced rocks and, if I'm lucky and the wind has been high enough to shake the trees, lots of pine cones for the fire. All this part, the first wood, the pines and the river gushing out from under the high, stone bridge, is deep and dark and secret, but suddenly the aspect opens out and what looks at first like a country park, is there before you, with one, great, glorious ash tree in just the right place.

It's an optical illusion, this sense of limitless space, for the meadow is not much above an acre and a half, with the river running down one side and bounded by another steep wood on the other, but the illusion never fails. I suspect that it's all got something to do with the exact positioning of that ash tree. Whatever it is, it has revived more flagging spirits than mine and, if I have a friend here who is tired or depressed or just generally fed-up, I despatch them to the meadow and never fail to see them come back with their troubles in perspective and a new spring in their step.

This lift in their spirits, and mine, may have something to do with the river which winds its way casually around the meadow, sometimes deep and slow, sometimes running lightly over beds of gravel and sometimes fast and furiously over great boulders, till it drops into the foaming pool under the second bridge. This bridge is not man-made like the other. It was formed, more years ago than anyone can remember, by a falling ash tree which left its roots embedded firmly in the bank and continued to live so full a life, that its top branches clung to the other bank and finally rooted in. Now it lies straight across the river and its middle branches have thrust up into trees in their own right, so that you can sit in mid-stream under their shade and watch the river swirling below. It's wise to make sure you're alone when you do this and not taking the dogs and goats for a walk, because they'll follow you blindly out onto the bridge and then declare that they can't possibly turn round again, so there you are, stranded on a tree trunk with a traffic jam of hysterical animals behind you. Then you must descend into the river, which is quite deep at that point, scramble onto the bank and gently extract them one by one from the rear.

The gentleman from the Plant Conservancy was mildly impressed by this natural bridge of mine, but not one jot by the strange little copse of straggly oak and alder and thorn standing on a curiously shaped mound nearby. Nothing else grows there, except moss. In autumn the acorns lie deep in the brown leaves, but normally the ground is quite bare under the trees. And yet I have a deep feeling about that copse; it seems to have a life quite separate from the rest of the meadow. I've wondered about it for a long time and thought I'd found the answer to its mystery in a book I discovered on the library van. It was an old book (and one I've failed to trace since) full of maps of long barrows and odd archaeological bits and pieces in this part of Wales. Our valley seems to be quite rich in long barrows, few, if any, explored. There were none precisely marked on my meadow, it's true, but perhaps they hadn't had a really good look.

'Nonsense!' said the plant man when I put this theory forward. 'You can see that the river used to run across the meadow here, till it was diverted for some reason. This is simply part of the old bank!'

I could see that he wouldn't be the sort of person who would believe that the dogs and goats are never very willing to stop in that copse but usually skirt round it distrustfully. Neither could he see, with the lushness of summer clothing everything, that just beyond the copse there are large grey boulders placed in what looks, to me, suspiciously like a circle.

We left the copse behind and went to examine a healthy patch of delicate feathery grasses which were gaily waving in the slight breeze, and then toiled back up through the woods to the farm, where Mrs P became tremendously excited about the wild angelica. Mrs P's latest project was herbs, tame or otherwise.

It had been a mini-project for quite some while, for Mrs P does like growing things. Not in gardens so much as in small receptacles like ancient teapots or redundant buckets or rusty tins with holes bored in the bottom. In spite of their unlovely containers, plants always grow well for my mother who chats away to them happily or sympathizes with their problems. And then Ann Young came to see us and the Great Herb Revolution began.

Ann, until then, had been the missing half of Graham Young, who bolstered up his own farm by doing odd jobs. He and Ann were living in an old barn, which they were frantically trying to make habitable, whilst coping with the inevitable dramas of their stock, and cultivating their organic vegetables which Ann sold at a local market along with her marmalade and exquisitely arranged bunches of herbs. Somehow, in between all this, Graham managed to fit in working for other people like me, who were in need of a bit of building or a hedge sorted out or simply slates put back on a roof. It was Graham who had endeavoured to build a goat Alcatraz for me the year before.

Tacked on to the main barn is a small yard enclosed by a stone wall containing a pair of very comfortable pig-sties. Blossom had long since decided that the pig-sties gave her claustrophobia and much preferred her shelter out on the field, so I'd used them as a summer residence for the goats. The trouble was that the kids could bounce out and over the stone wall without the slightest trouble, so I'd asked Graham to extend it to a goat-proof height. He laboured doggedly all day and then declared that nothing could possibly scale that wall!

'Want to bet?' I said cynically and went to fetch Dolores' daughter Loveday, an ethereal creature with incredibly long legs and a dim but determined nature. I put her in the pig-sty and stood back. She didn't even bother to size up the distance, just made one almighty leap and landed safe and still running, on the other side of Graham's wall.

As we watched her disappearing down the field in pursuit of her mother and sister, Graham set his mouth grimly and went to collect more stones. His wall grew and grew until finally it did defeat the kids, who spent a frustrated few weeks behind it while I was weaning them from their mothers.

Of course, long since then they'd all grown up and graduated back to the main goat-shed, but if anyone was bent on causing mayhem, back they went to Alcatraz until they cooled down and, later on, it was to come in very handy for another large resident.

Meanwhile, Graham Young had reappeared whenever I was in need of help. Often when he came, he would proffer some of Ann's marmalade or a loaf of her home-made bread or some vegetables from her garden, but of Ann herself, we never saw anything. Until one day she arrived with her lean blond husband, a round, rosy-cheeked girl carrying a large, old-fashioned basket laden with herb plants, bread and a most delicious quiche, and Mrs P took her to her heart.

Those original herb plants which Ann gave Mrs P, and which were to spark off the new obsession, are burnt into my brain. There was comfrey, lemon balm, apple mint, lovage, wormwood and pennyroyal.

It was the pennyroyal, otherwise known as 'lurk-in-the-ditch' or 'pudding grass' (because it was used to flavour them), which immediately excited Mrs P.

'That's the stuff,' she said to me when Ann and Graham had left, 'that my own mother used to make me go and buy a pennyworth of from the chemist when I was a little girl!'

'From the chemist?' I asked.

'Yes. You bought it dried, in little twists of paper. My mother was very secretive about it and told me to go and get it when my father wasn't around. I always wondered why!'

Hastily we consulted the copy of Culpeper's *Complete Herbal* and there, under 'Pennyroyal', it said, amongst a lot of other things:

'If boiled and drank, it provokes women's courses and expels the dead child and afterbirth.'

Mrs P and I gazed at each other in wonderment.

'That explains it,' she said. 'Poor love, with a husband like Pop, I suppose she'd try anything. Can't have worked too well, because she had ten children. But you know that's answered a question which has always bothered me!' She took the Culpeper and was not heard about the house for a good few hours as she sat absorbed by all the amazing properties the master claimed for his herbs.

The next day I was set to find suitable sites for the herb plants Ann Young had given us and from then on, Mrs P and Winston were to be seen out with the herbs every day, watching over them and willing them to grow big and strong. Thus encouraged, they burst forth in profusion and from then on we ate and drank herbs to distraction.

I must say the pennyroyal is delicious in salads (although we did warn any young, fecund visitors to be wary of it), and I'm never without a goodly supply of it at lambing and kidding time to give to the newly delivered mothers to help with their cleansing. Whether it really works or not (and I've read since that old cow-men and shepherds fervently believed it did) the sheep and goats wolf it down even more enthusiastically than the ivy leaves which are supposed to perform the same function of making the womb contract.

As for the comfrey, it really is a miracle herb when it comes to stopping a wound bleeding. A few bruised, fresh leaves and the effect is almost instantaneous. And, whether they fancy it or not, every single animal on the place gets comfrey in one form or another, straight down the hatch, when they're off colour, because its healing properties are almost as good inside as out.

Thus far I accepted Mrs P's new enthusiasm and even added to it by buying more herb plants for her, but it did have severe drawbacks. I learnt to duck fast when I saw her coming at me with yet another steaming bowl of her more varied concoctions and, although now she is a little more circumspect and accepts that some herbs do not taste at all nice, even when flavoured with enormous dollops of honey, in those first frantic days, she stood over me while I drank up truly awful combinations of herbs, looking, as she said, for something to calm my nerves and make me a nicer person to be with.

Actually, my nerves were in a pretty bad state. I was anxiously waiting to hear if the pages of notes I'd sent to Michael Bowen had met with his approval and I could go ahead with the actual pilot programme; in spite of the odd golden day, most of the summer had been generally foggy and damp and I was despairing of ever getting the hay made and, as autumn began to creep upon us, there was the not too distant prospect of getting five goats mated.

There was nothing I could do about the pilot except plug Mrs P into her Hot Line and wait. The hay did get made at the beginning of

September, when the sun finally managed to break through for several days at a stretch.

Myrddin, in company with everyone else in the valley, was desperately trying to get the last of his own hay in, but he still found time to put his tractor furiously at my one field and get it cut and turned in record time while I danced up and down and got in his way. He was quickly followed by Cynog Davies from the farm at the top of the hill, who owned the baler and soon, the flat, green swathes of grass were gathered up and there, shining and compact, were nearly a hundred bales of hay.

Clouds were scudding up from the west as I heaved and pushed and rolled those bales down the hill. Even Mrs P came out and gave a few tentative pokes at one or two, till there was a shout from the gate and Myrddin came back with Gwynneth and her husband Eddie and helped to fling the last few bales up into shelter as the valley went a deep, violet-tinged green and the swallows swirled over the mown field in one final salute before the rain came back and broke the truce.

We grinned at each other, all of us with our hay safe in at last and knowing that in spite of it being so late in the year, it was good, sweet hay, and the weeks of agonized weather-watching were over. Still glowing with achievement, I went inside to ring Christine Palmer and see if she had any solution to the goat problem.

Chapter 9

Eighteen months had now passed since the two goats, bloody-minded Nana and elegant, devoted Dolores, had given birth to twin female kids apiece. Nana, whose hatred of the human race was really centred on me because she associated me with the initial separation from her own adored mother, had been sold into a life of luxury elsewhere and Dolores reigned supreme over the little herd.

There had been a few dicey moments when Dolores came out of sick-bay after her disastrous second kidding, but it took her precisely half an hour to put down the rebellion, utterly and irrevocably. The little male I'd salvaged from that kidding had gone to live with my friends Meta and Alan Bonney, and in time became one of their stud males, and even got his photograph in the little book Meta wrote about goats.

Meta had looked after my own farm in the traumatic months when I was transferring from London to Wales for good, and had come back briefly to help me just after Mrs P left for Australia, but she and Alan had since moved further away to Cardiganshire to grow potatoes, along with their goat-keeping. Fortunately for Dolores' son, they didn't bother about show potential in their goats, but concentrated on good milk production, and that was something he could pass on to his offspring, for his own mother's milk was rich and profuse and his father was a very grand goat indeed, called Chessetts Concorde.

Conky was standing at stud with Christine Palmer, who had sold me Dolores herself. He was getting on a bit by now and, while I still had the chance, I wanted to use him for the third time on Dolores. The other four goats were his own daughters and would have to be served by some other male.

'Well, there's Cavalier and Chorister,' said Christine Palmer when I rang her. 'Cavalier, of course, is a Toggenberg.'

'I thought they were all Toggenbergs,' I said.

'No,' Christine explained patiently. 'Conky and Chorister are BRITISH Toggenbergs. The Toggenbergs are the pure Swiss line, unmixed with British blood. They're a bit smaller and hairier than the BT's. Their milk production isn't usually quite as high, but they're very good at milking through without having to be put in kid every year. Now, as Dolores has got a bit of Anglo-Nubian blood in her, a breed not usually renowned for milking through and her colour is certainly not proper BT, then it might be as well to put her daughters to Cavalier. Nana's two can go to Chorister if you like.'

'Right,' I said, a bit muddled by all this science. 'The trouble is, I don't want to make five journeys over to you.' For Christine had also moved, all the way to Pembrokeshire, and the round journey would take the best part of a day. 'The idea of travelling with even one of them, hysterically in season, on the back seat of the Beetle, is too awful! You know what Dolores is like? Well, the others are just as bad. I'd have to find someone with a trailer to bring us and that would be very expensive five times round!'

Christine knew only too well what Dolores was like. Two years ago, when she'd boarded her and Nana at mating time, her well-run goat herd had been turned on its collective ear and tail. The previous year, Dolores alone had been delivered, served and taken quickly away, to the intense relief of both Christine and Conky, for even he found an amorous Dolores on the premises a bit much.

There was a long silence on the end of the phone and then a sigh.

'I suppose you *could* leave them here again till they've all been served,' said Christine in a martyred voice. 'I'm afraid they'll have to live in a horse box I've got on the yard though. I really couldn't go through with them disrupting the place again. Not five of them!'

'Well, four actually,' I soothed her, 'because what I plan to do is wait until Dolores starts calling and then bring all five of them over together. Conky can serve Dolores and I'll bring her back with me. At least you won't have her to worry about!'

'That,' said Christine fervently, 'would be much better.'

So I arranged with Myrddin that as soon as I saw Dolores flying up and down the goatshed, terrorizing everybody, including me, in a total change from her normal, intelligent self, and glaring at the world with mad, yellow eyes and possessed of the strength of demons, I would ring him and he would collect us all in his Landrover and trailer.

Even so, the journey was fairly grim. We had to tie the goats up in the trailer because the younger ones would have had no chance of escape from Dolores, who was truly frightening when she was in season; leaping passionately on the others, rearing to her full height to bring her head crashing down on anyone who got in her way, biting ears and tails, and all the time keeping up a raucous bleating which was ten times her usual volume and totally deafening. Things were even more complicated by the fact that her daughter, Minnie, who is the only other goat I've known to behave in quite such an abandoned way and be possessed of so much insane, lustful strength, also chose that morning to come into season. Of course, in one way it was a good thing, because it meant that we could bring two goats back with us and only leave Christine three to cope with, but it didn't make for a very easy journey for the rest of the goats with two maniacs on the rampage in a confined space. Even Whacky, Nana's daughter, who could be a very nasty bit of work when she wanted, was desperately trying to get away from them instead of giving as good as she got.

'That should hold them!' said Myrddin grimly as he finished the last of his never-to-be-got-out-of ties, gave it a final tug and leapt out of the way as Dolores tried to give him a passionate bite. 'We'll stop every few miles to see if they're all right.'

It was a long and weary way to Clarbeston and the final lurching jaunt down the long, pot-holed track to Christine's new small-holding, didn't help our nerves too much either, although it did shut Dolores and Minnie up for a bit.

There was a deep, blissful silence as Myrddin finally switched off the engine of the Landrover and Christine emerged from her cottage.

'They're very quiet,' she called. 'Not like Dolores to be so quiet!'

We let down the back of the trailer carefully and saw the five goats standing there queasily, ears down and a slight shiver running across their backs.

'Ah!' said Christine, 'the track! It's terrible, but it doesn't belong to us and the man who owns it won't do anything about it. You should see it when it's raining. Those potholes are like lakes. Anyway, it's cooled Dolores off!'

A loud roar came from one of the nearby sheds. Dolores pricked up her ears and sniffed rapturously as the scent from the yard reached her. Suddenly, Myrddin's miracle knots were as nothing to Dolores. With a flick and a twist, she shed her bonds and was out and dashing across the yard, her tail wagging frantically, calling as she went.

'That awful goat,' cried Christine and rushed after her. 'Poor old Conky won't stand the strain,' and she fielded my little passion-flower just as she reached the shed where Conky was leaning his great head hopefully over the door.

'You just wait my lady!' Christine stormed at Dolores. 'This time you're going to be civilized about it,' and she frog-marched her away to another shed so that we could unload the rest of the goats in peace.

Half an hour later, Whacky, Loveday and Little-Nana were settled into their temporary home in the horse-box with hay and water. Minnie had been served by Cavalier in a fairly cool fashion, Christine and I were recovering from trying to control Dolores while Conky got on with his job, and Myrddin had strolled off for a walk, unwilling to run the risk of his clothes being perfumed by the heady smell of male goat, for we had another visit to make that day.

'I've got a present for you, by the way,' I told Christine when Dolores and Minnie, their lust slightly abated, had been shut back in the trailer. 'You know you said you were looking for a rooster? Well, I've got a beauty for you. He's the son of my Poland rooster and dear old Elizabeth, the Rhode Island hen who lays such massive eggs. He's big and resplendent and he's inherited some of his father's crest. As long as he doesn't frighten your hens off, he'll be very decorative about the place.'

I led the way round to the front of the Landrover where the rooster was still waiting quietly in his box. Carefully, I opened the lid.

'Look,' I whispered, 'isn't he absolute bliss?' (I tend to talk like that when I'm not too sure of my ground.)

Christine peered down at the rooster which was a lovely, deep, russet colour with beetle-green flashing on his wings and a spike of black and gold feathers on his head.

'Yes, he is,' she said, 'and that's what I'll call him! Absolute Bliss!'

'Well,' I said defensively, 'he *is* good looking! But I'd keep him securely locked in for a few days. He might take after his father in more than looks.'

'Ah!' said Christine. 'That's the one you can never get in at night is it?'

'Yes, but Elizabeth is a dear old hen and he must have some of her in him too, so just keep him locked up till he gets used to you and then I'm sure he'll be as good as gold.'

Christine looked at me uncertainly. 'We can but try,' she said at last and picked up the box gingerly and carried it across the yard.

'Are you ready then?' asked Myrddin as he strolled back from his walk.

'All ready,' I replied. 'You back the trailer round while I just go and say good-bye to Whacky and Loveday and Little-Nana.'

'Don't be long,' he warned, 'we've still got a long way to go.' For he had discovered, by consulting his eternal maps, that we were within striking distance of yet more friends who had moved to Pembrokeshire and indeed, I'd used them as bait to get him to drive the goats and me to Christine's place.

Gerald and Imogen Summers had been directly responsible for my move to Wales when, at the beginning of the seventies, I'd come to interview Gerald, at the Black Mountain Bird Garden, about his first book *The Lure of the Falcon*. He and Imogen lived a hectic life in a little witch's cottage surrounded by mounting vegetation, twenty dogs, a cacophony of small birds, and Random, Gerald's tame golden eagle. The Artist, who had done the illustrations for the book, and I had spent an hilarious but enchanted time with them and, with our brains softened by the experience, had fallen in love with the tiny farm in a nearby valley, which I now call home. For years, when things got a bit out of hand, or we were feeling overcome with our latest rural nightmare, the Artist and I had rushed to Imogen and Gerald to be re-enthused and regaled with whisky and generally put back on our feet again. After the Artist had left, I had still found the Summers and their precarious, but never dull, existence, enough to cheer me up when I was low. And then they, too, had gone to Pembrokeshire.

The last time I'd seen them before their move, they'd just rescued another dog which nobody wanted. The whippets and lurchers and their big mastiff had been banished to the rest of the cottage and many of them took their repose on the big, double bed upstairs. Others whined anxiously at the door of the living room, which Imogen always managed to make cosy and chintzy, in spite of living with Gerald to whom furniture was just another perch for a bit of itinerant wildlife.

The new dog was extremely fierce owing, no doubt, to having had the tops of its paws removed in the interests of some obscene science, and thus having to get about in little leather boots. Imogen was keeping it in isolation in the living room, which had once been a barn tacked onto the cottage.

I'd arrived with a friend to let the Summers hear a tape I'd made some weeks before of Gerald and Random; a pair of peacocks mating on his foot whilst he was in full stream about his beloved eagle; Imogen's own remarks about their life punctuated by various rude remarks from their parrots and, above all, the famous dog 'aria' which their pack would suddenly break into for no reason at all and which, until then, nobody had ever succeeded in recording.

It was a warm, sultry day and the top half of the split door, at one end of the room, was open on to the yard, to let the fresh air in and the barking of the new dog out. Imogen greeted us warily and told us to sit very still in case the dog bit us and she sat beside it, feeding it her best cushions to pacify it. Gerald was walking up and down telling us about the last time Random had escaped and terrorized the neighbourhood and how he'd been up all night looking for her, while Imogen politely called pleasantries to my friend above the sound of the dog growling and swearing into its cushion. After several attempts, I got everyone, except the dog, to be quiet and switched on the tape recorder.

We were about five minutes into the tape when the room went dark, and an almighty burst of thunder smashed from the sky and brought torrential rain down with it. The dog went completely berserk and howled dreadfully. Imogen patted it absent-mindedly, and fed it a bright scarlet cushion. My friend sat with a fixed smile on his face, eyes staring. There was more confusion at the door and three faces, encircled by an enormous black umbrella, on which the rain was rattling furiously, appeared over the top half of the door which was still open.

Gerald bounced forward to see who they were and bellowed back the information that they were from some wild life park in Devon and could Imogen rustle them up a cup of tea or something?

'I can't,' his wife cried above the racket. 'They can't even come in because the dog is really mad now. Do sit quite still Jeanine and, er, I didn't quite catch your name . . .' and she smiled brightly at my friend.

'Sorry,' Gerald roared at his visitors. 'Wife says the dog's not safe. Can't invite you in.'

'But we've come all the way from Devon,' wailed the three faces at the door.

'Yes, and it's lovely to see you,' shrieked Imogen, 'but you can see how it is with this poor dog!' It would never have occurred to her or Gerald to inconvenience the dog even for the most illustrious visitor.

What with the noise of the rain, Gerald's thunderous conversation with his bedraggled visitors at the door, which he conducted as if nothing was amiss, the dog's enraged barking and hysterical comments from the two parrots hanging from the ceiling, Imogen was finding her duties as a hostess confusing. She patted the dog soothingly, she constantly apologized to the mournful visitors, she swore back at the parrots, she asked my silent friend what he'd just said and finally she heard with relief someone making a remark which she could understand and with which she was in full sympathy.

She turned eagerly to the voice. 'Oh!' she cried, 'I do *so* agree! You're absolutely right!'

The dog nearly had my hand off then because I went into convulsions. The voice was Imogen's own and it had emerged out of the tape recorder which had been burbling quietly along on its own all the time.

I missed the Summers terribly when they went, but Imogen was delighted with the move.

'We never got a minute's peace you know, with all the visitors to the Bird Garden,' she told me, 'and now we're right near the beach and the cliffs and Gerald can fly Random up there and it's a very neat little house with everything in working order. There's even a great big sun-room where we can sit and get right away from all the dogs. I've told Gerald, no birds, no polecats, no dogs, just people in that sun-room!'

As Myrddin and I approached their new home, I wondered how they were getting on with only a tiny collection of birds and a much reduced dog population, owing to a virus which had decimated the pack when most of them had been in kennels while the Summers were moving. It almost broke Imogen's heart, especially as her adored mastiff, Martha, had been one of the casualties.

I was telling all this to Myrddin as we drove along.

'It's probably just as well though,' said Myrddin wisely. 'Just a couple of dogs will be quite enough for them in a much smaller place. It's a pity about the birds though. At least he's still got Random!', and he smiled happily for Myrddin is a great admirer of Gerald and his talent as a bird handler.

We arrived at the little bungalow with its green roof and white walls and orderly garden, high up above the village of Solva, and I felt rather sad. It didn't look at all like the Summers. And then Imogen bustled out of the door and Gerald's booming voice could be heard shouting, 'Just in time for a whisky,' and somewhere, what sounded suspiciously like more than a couple of dogs, were barking furiously.

We parked the Landrover and trailer carefully, checked that Minnie and Dolores were all right and fought a rear guard action with Imogen who wanted to release them into the garden. Finally I distracted her by some pots of yoghurt and cheese I'd bought from Christine and we went in to the new house.

I needn't have worried. There were the parrots, still swearing; there were assorted cages and perches around the room, all fully occupied; there was the muddle of falconry bits and pieces which followed Gerald everywhere and there was a large and utterly spoilt pack of dogs occupying every available seat.

'And here's Hannah,' said Imogen, 'my new mastiff,' as a large brindled creature lumbered in. 'Most of the others have come from the Dog Rescue. I mean we couldn't refuse could we, not when we had so many vacancies!'

A deep suspicion assailed me. 'I'm longing,' I said, 'to see the famous sun-room.'

Imogen gave me a furtive look. 'Ah yes! The sun-room!', she muttered and led me round the outside of the house. I peered in.

It would have been a perfect place for sitting calmly in the sun, that spacious, airy room with vast windows placed in just the right spot. But there were no sun-lounges, no wicker tables with drinks reposing on them. Instead there were the remains of several well-chewed carpets, a couple of extremely unhappy sofas with most of their stuffing gone and there were two fierce little faces glaring at us from behind the glass.

'What,' I demanded, 'are THOSE?'

'Those?' said Imogen innocently. 'Those are Gerald's pugs! They're only puppies so we had to let them have the sun-room to run about in. Gerald's very fond of pugs, but I'm afraid so is Random. So you see it's safer for them in here than in the garden!'

Eventually, Myrddin and I made our way rather sadly back to the valley and, as we passed their old cottage, now in a perfect frenzy of renovation under its new owners, both of us looked at it and sighed. Things would never be quite the same without the Summers.

It was three weeks later when Minnie informed me that she needed urgently to go back to Clarbeston.

'Oh no,' I told her. 'Do you mean to say all that cavorting around with Cavalier didn't do any good?'

Oh it had done some good, the evil little gleam in her eye assured me, but it hadn't got her in kid.

Desperately I rang Christine Palmer and gave her the news.

'You'd better get her back quickly then,' she said. 'How are you off for transport, because you can take the others back with you now? They've all been served.'

'There's not a chance of Myrddin taking us,' I moaned. 'He's off somewhere else today.' I glanced out of the window and noticed the Young's ancient Datsun estate just nosing into the yard. Graham had come to do some fencing for me.

'There's just a chance,' I murmured into the phone. 'I'll ring you back.'

'No problem,' said Graham when I asked if his estate would fit four large goats in it. 'Be a bit of a squash mind, but if you're game, I am.'

Shortly, we were on our way with Minnie, firmly tied up in the back of the Datsun, breathing passionately down our necks. When we arrived at the Palmer's, Loveday, Whacky and Little-Nana greeted me calmly with an air of having learnt a thing or two about Life. Minnie was served again by Cavalier while Conky looked over his shed door mournfully and then all four goats were compelled back into the Datsun by Christine. They just fitted although, looking at the photo I took of them peering through the windows, it seems as if there were about twenty goats in there.

'Oh! How is Absolute Bliss?' I asked Christine as we were about to drive off.

'Absolute Bliss,' she said firmly, 'has gone!'

'Gone? Where?'

'Over the hills and far away. Well not so far really, but far enough. I kept him locked up, as you said, for three days. On the fourth, I was just opening his door a crack to feed him, when he shot out past me and took off straight up the track. He didn't make a sound, but I've never seen anything move so fast. He had his wings out slightly to give him more push, his legs went like clockwork, sort of sideways, his head was down and he leapt those pot-holes like a showjumper! He looked just like that roadrunner character with his spikey head! I sent the old sheepdog after him but he didn't have a hope of catching him! Anyway, he's settled down in the little wood over the hill there. We've tried and tried to catch him, but it's hopeless. I'm sorry. I suppose a fox will get him eventually.'

But it didn't. Absolute Bliss was still living a full and happy life, surrounded by a quartet of adoring lady pheasants two years later in his woodland setting. As Christine said:

'Absolute Bliss is what he was and Absolute Bliss is what he's got!'

Chapter 10

The rain streamed across the deserted car park and caught me full in the face as I got out of the car, slammed the door shut, and rushed away down the hill where everyone else had parked.

People were milling around the open doors of the auction rooms, trying to push their way in out of the wet and being frustrated by those sensible enough to get there early. Ignoring the press of bodies behind them, they lingered over the sad array of defunct kitchen units, piles of ragged, dusty books tied up in bundles, rickety tables, and shiny dressers which were a long way from the valuable old oak ones the dealers were hoping for. Neither did anyone believe the labels 'In Good Working Order', hopefully decorating the ancient TV sets and sewing machines, although in the heat of the auction doubtless someone would suspend their disbelief. The more valuable items were much further within the room and there, the rows of chairs, likewise for sale, had already been firmly occupied by the regulars.

I pushed my way carefully amongst the damp spectators and hovered anxiously near the auctioneer's rostrum and tried to catch his eye, while at the same time endeavouring to escape the notice of the crowd. It's something I've never been able to overcome, this embarrassment when I'm recording background in a crowd; standing there like a complete fool, holding the microphone up like a banner and twiddling with the controls on the machine. Once I've got someone specific to aim the microphone at, then I forget how unreal the situation is and become absorbed in the interview, but to stand there, recording plain sound, shifting from foot to foot and never knowing what to do with my face, is something I do a lot to avoid. This time I was completely out of luck.

'Do you think you can shift up for this lady, please,' the auctioneer shouted at the rows of chairs in front of him. 'She's from the BBC and she wants to record the auction!'

Frantically, I tried to shut him up and indicate by sign language that I'd much rather stand anyway, but by then a lot of shuffling and shifting was going on and, before I could protest, I was manhandled into a tiny space between two large ladies, delighted with this diversion to occupy them till the auction started. I sat there, elbows wedged in, tape recorder perched precariously on my knees, the microphone cord suddenly living a life of its own and snaking onto the floor out of reach, and the mike itself within inches of the heavy breathing coming from my two sponsors.

'Are you from the telly?' one whispered loudly, while a row of heads turned to hear the answer.

'Er . . . no. Radio.'

'And when can we hear it then?'

'Um . . . I'm not sure whether it will be on the air at all,' I floundered. How to explain, I wondered, that this was a pilot programme with a very uncertain future and why did I find it necessary to explain anyway? Why not just tell them anything? I cursed Mrs P once again for bringing me up to be as honest as I could and my father's genes which were precise and pedantic to a fault. Telling the truth has landed me in a lot of trouble in my time. Now, my companions looked at me suspiciously. I was obviously there under false pretences.

But when the auctioneer's gavel sounded above our heads for

silence and I depressed the keys of the tape recorder to set the tape whirring around, both ladies leaned over to look and one shrieked joyously:

'Better be careful what you say Gladys, or it'll be taken down and used in evidence!'

'Yes, and you watch your swearing!' hooted Gladys. I sighed and surreptitiously pressed the pause button.

By now everyone in the front row was staring at me and throwing remarks at Gladys and her friend. I sat and got hot under the collar and prayed for deliverance. It was going to take a while, I thought sadly, for me to get back into the way of all the old tricks of the trade. Once, caught in this situation I'd have been adjusting the levels on the machine like mad to record all the remarks flying around me. Finally the auctioneer, grinning all over his face, put me out of my misery.

'Have you quite done ladies?' he called, 'because I'd like to have a sale here sometime this afternoon if you don't mind! Right! Lot Number One!'

I made up my mind in a hurry. 'Excuse me,' I murmured to my new friends and, groping around their legs for the mike lead and holding all my recording bits and pieces aloft, I crawled across their knees and escaped.

The auctioneer raised his eyebrows at me, indicated that he'd meet me later for his interview and I went to stand well back to record the few moments of actuality I needed to back up that interview. It was to be about the chances of picking up enough bargains, at a saleroom in a small market town, to furnish a cottage if you happened to be pretty broke (as I had been) when you made your great leap into the country.

Eight years before, I'd attended the very same rooms and the very same auctioneer had knocked down to me, marvellous things that cost practically nothing then and still form the bulk of the furniture in my own cottage. The sort of things which were either absent now or, if they were present, going for fabulous prices. I thought smugly of the big grandfather chairs I'd picked up for a couple of pounds and looked at one, not half as good, that was being hotly bid for at something approaching a hundred.

Times had changed a lot, but those shiny, glassed-in dressers and the tatty 1940s wardrobes and cupboards could look quite good with a lick of imaginative paint and it was bargains like that I wanted to tell my listeners about. For I wanted to make my programme quite different from the glossy pictures in the magazines, where country cottages were stuffed full of antiques and there wasn't a stray bit of mud or straw to be seen. Years later, someone was to describe it as 'a programme for people with real mud on their wellies', and it was the best compliment I ever had.

This interview, of course, was only one of many I was hastily collecting for Michael Bowen, who had at last rung to say that the pilot had been commissioned and could I have it ready to record at the beginning of December?

That November I ignored the torrential rain which blessed us, the gales that shot every last withered leaf from the trees, and even the heavy snow which came early and had all the old hands shaking their heads and predicting a dreadful winter ahead. I drove around interviewing people, and then came home to bend over my ancient editing machine and try to re-learn how to cut half an hour of rather tedious tape into a riveting two minutes with sound effects all flowing easily behind. Apart from remembering not to cut a sheep off in mid-bleat, or make a cockerel sound as if it was being swiftly strangled, there were indoor hazards like ticking clocks which had to have all their beats matched perfectly instead of having a mad fit of hiccups in the background.

I noticed idly, whilst listening to the tapes, that my old editing machine had wood-worm in its casing and wondered if it had acquired enough antique status to get the dealers fighting over it yet. It had been at least twelve years old when I'd bought it ten years before and working its massive controls was like driving a tank, but together we got the tapes in order, I settled down to write a thrustful script and, on the appointed day, turned up in Bristol.

'I don't think much of the name you've chosen for it,' said Michael. 'A Small Country Living . . . sounds a bit religious.'

'Well, it's meant to be rather a play on words,' I said. 'But I've got about another twenty titles here and this seemed to be the best.'

Michael scanned down the list, wrote in a few of his own and consulted the two Studio Managers. Half an hour later we all agreed on 'A Small Country Living' and that's how I announced the pilot and that's what we wrote on the tape box which was to go to London for the Controller's ears.

My own ears, when I returned home, were once again assailed by the tearing gales, and I was too busy getting the animals' sheds ready with deep beds of straw, or carrying in the endless loads of wood and coal, or generally battening down the hatches, to give much thought to whether the pilot would blossom into a full-blown programme. Meanwhile, Christmas was creeping up on us and there were enticing smells seeping out of the kitchen, frantic attempts to get cards and parcels done in time, and a tree to be found.

'Do we really want a tree?' I asked Mrs P wearily.

'Well, I suppose not, if it's too much trouble,' she replied stoically. I looked at her carefully and noticed the droop in her shoulders and the dimming of the excitement in her eyes. To Mrs P, the heart of Christmas is the tree. Apart from the fact that she believes completely that if you wish on the tree on Twelfth Night, your wish is sure to be granted, she loves glittering, sparkly things and the tree is a perfect excuse to use them lavishly.

'Trees are terribly expensive this year,' I told her, 'and from what I've seen of them, they're very straggly and dry. What I'll do is get an old branch from the woods and paint it white. That'll look very effective with the decorations on it.' But I made up my mind that I would go into town the next day and buy her a real Christmas tree.

I was woken early next morning by a still brightness glowing through the yellow blinds in my room. Pulling them up, I saw the valley deep in snow. It had come without a breath of wind, so that it lay straight and smooth, and every tree, gate and fence was delicately feathered in white.

The snow was obviously too deep for my decrepit Beetle to get me into town. So, as soon as the stock were fed and watered, I announced to the dogs that we were off to the woods to look for fallen branches. As the front door opened, the whippets stepped gingerly out into the white world, before jinking and twisting and flinging the snow up in clouds about them. There was a horrified cry from Mrs P behind me as Winston leapt from her arms and rushed out to join them.

'He'll be buried,' she wailed.

Winston was nine months old by now and, although it had

become obvious that he would indeed survive the basic rigours of the cottage and bore no resemblance to the delicate little chihuahua Mrs P had longed for, he was still small enough to make the dogs' water bowl look like his own private swimming pool. As he launched himself, therefore, at the snow, he disappeared completely and Mrs P's lamentations rose on the air. A jet of snow burst upwards beside her and Winston reappeared, a joyful grin on his fat bear's face, before he disappeared again and the only sign of his going was a rapidly moving hump of snow. He re-emerged, round and white and immensely pleased with himself. His fury when I picked him up expressed itself in a miniature tiger's snarl, his little nose wrinkled, tiny teeth bared. Unhesitatingly, he sank them into my thumb and drew blood.

By the time Mrs P had dug into her store of dried comfrey and made me a poultice to stop the bleeding and Winston had been towelled dry, to his intense indignation, and everything had calmed down again, I'd forgotten all about the expedition to the woods. Late that afternoon, when all that was left of the wintry sun was a faint blush beamed across the valley onto the snow-covered Fans, a Landrover churned its way into the yard. Behind the wheel was the friend who had driven Mrs P to London the year before to catch her plane.

'I've been giving the boys in the forestry a bit of a hand,' he said as he climbed down, 'and they said I could have a few trees. Have you got one yet, because I've got one going spare.' And from the back of the Landrover, he hauled a perfect little Christmas tree.

That night, while Mrs P surpassed herself making a record number of mince pies, attended by Winston on guard by the Rayburn in case the whippets forgot themselves as they drooled at her from the door, I brought out a large box of very old friends. They were blue and green and old-rose silvered balls, long swathes of sea-green tinsel and three small, rose and gold angels.

I had bought these decorations when I had my first Christmas in my very own flat and, except for one or two casualties, the same collection has graced every tree since. The predominant effect is for the blue and green baubles to do little more than enhance the deep green of the tree itself and, winking out from within the branches, comes the hint of old-rose. The three angels descend from the top of the tree and there is nothing else to distract from its shape; no hideous electric lights blinking on and off and trailing their wires everywhere. The tree stands against the grey, stone wall at the back of the room and an old converted candle lamp (once used to light a stage coach) sparks lights from the coloured balls as they move very slightly in the draught when the door is opened. Beneath the tree stand two zany Staffordshire lions, grinning in welcome at everyone who comes in.

Some people may find it boring to repeat, year after year, this undramatic tree decoration, but I find it very comforting, as old familiar things are, and it reminds me of all the people who have shared Christmas with me under trees that looked very much the same.

Something else which comes into its own at this time of year, is the incense burner in which I burn, on little charcoal discs, sprinklings of very special incense. The burner itself is in the shape of a ki-lin, the oriental version of the unicorn. They are not dissimilar animals, although the ki-lin's horn sweeps backwards and, as Odell Shepherd says in his book *The Lore of the Unicorn*, the ki-lin is not white.

'He is resplendent in the five sacred colours which are the symbols of his perfection.'

Earlier, Mr Shepherd tells us: 'Chinese writers do not assert that the unicorn or ki-lin is a native of their land; on the contrary, they say it comes from afar, presumably from heaven, and only at long intervals of time. They regard it, so to speak, as an intermittent animal, and its appearance on earth is considered a certain omen of a beneficent reign or of the birth of some great man. . . . To say of any man that a ki-lin appeared at the time of his birth is the highest form of flattery.'

Odell Shepherd's book was one of the first things I bought when I began my collection of unicorns and that reference to the ki-lin was the only one I could find for years. It's hard enough to find replicas of the European unicorn, so I gave up any hope of ever seeing a ki-lin in the many antique shops and markets I haunted. If I did ask them, the dealers were utterly bewildered, and even Jack and Muriel Sassoon, with their many contacts in the oriental art and antique worlds, told me the best I could hope for was to look at the few ki-lins in museums.

103

As so often happens when you're looking for something, my ki-lin turned up when my mind was concentrated elsewhere. On the actor Peter Bull's teddy bears to be exact.

For some reason, it was quite early in the morning when I'd climbed the dark stairs to Peter's little flat in Chelsea and partook of coffee with him and the collection of bears about which he'd written in his book *Bear With Me*. As usual, I was there to interview him about that book and, by the time I'd met all his familiars and we'd talked about bears ancient and modern for a long while, my head was full of bears and very little else. As I descended into the street again, I noticed the antique market along the way, stirring into life as the stallholders arrived, undid the grilles in front of their cubicles, and set out their wares for the day.

There was no one at the cubicle where I stood riveted to the spot, gazing in sudden hope at the little seated figure on a shelf at the front. It was black with age, shabby and broken, but its purity of form, and quiet dignity, shone from the jumble of Victorian and Edwardian junk all around.

The look on my face when I collared a nearby stall owner was everything it shouldn't be when you are hoping to pick up a bargain, but it made her say hastily, 'Don't go away. I'll get him for you!'

When he arrived ten minutes later, he was thin and wiry and long-haired and utterly amazed when I told him what I wanted. Like me, he wasn't really fitted for this dealing business, because he blurted out:

'That's been here for ages and you're the first person who's ever shown the least bit of interest in it. It's a unicorn.'

'A ki-lin,' I interrupted him.

'I don't know about that,' said the dealer, 'but I did take it to the V & A, and they told me it was a seventeenth-century Tibetan incense burner. Look, I've written it on the label underneath. You can check with them if you like. It's bronze.'

My heart sank. If it was so old, I certainly wouldn't be able to afford it. It rose again when he told me the price. It might have meant a bit of belt-tightening, but £20 was well within my range. I had no intention of going near the V & A with it either. Much better keep my illusions and the ki-lin didn't need a pedigree to make it beautiful in my eyes.

As an incense burner, the ki-lin was a bit of a disappointment at first. Its body is hollow and a small elliptical piece of bronze lifts out of the back. Its mouth is slightly open. I gathered that the incense went into the hollow and the incense then rose from the ki-lin's mouth, but when I tried burning little pieces of joss stick inside it, they died out at once. Besides which I don't like the cloying smell of joss sticks. I gave up the idea of using the ki-lin as an incense burner and just placed it on a carved, wooden, Chinese plinth which Muriel found for me, and enjoyed its beauty. It was Muriel, though, who also solved the incense problem.

'Why don't you try the Astrological Emporium round the corner?' she asked one day. 'I was there this morning and they've got small glass jars of incense and it's got a lovely smell.'

When I entered the little shop with its clanging bell, there was Peter Bull himself, greeting me with a bow amongst the wild assortment of astrological books and the glass, china and jewellery all bearing the various signs of the zodiac.

Peter and his friend Don had opened the shop not long before and, while Don got on with the ordering and the day to day business, Peter had a lovely time hamming it up by welcoming his customers with a flourish and bowing them out again, booming, 'Thank you for your esteemed custom Modom.'

'Incense, Modom,' he said now as he handed me into a chair by the counter. 'My esteemed colleague will be only too happy to oblige.'

Don raised his eyebrows at me and then brought forth a row of glass jars marked variously 'Sun', 'Moon', 'Venus', 'Mars', 'Elvin', and 'Astar'. With each was a roll of silver paper containing round discs of charcoal.

'You light the charcoal on something like a saucer,' said Don and, as he demonstrated, the charcoal fizzed and sputtered and gradually settled into a steady, red glow. 'Now, you sprinkle the incense on top and hey presto,' and he stood back to see my reaction as a plume of pale smoke rose in the air and the shop was filled with a strong, woodland smell.

'What is it made of?' I asked him, sniffing deeply.

'Depends which one you use, but basically they're made of tree gums and seeds and herbs and things and they're sort of 'brewed up' by a group who live out in the country . . .' he paused and looked at me, 'Er . . . when the stars are in the right conjunction for that particular incense.'

'I don't believe it!'

'Oh yes, because sometimes we run out of one of them (they're very popular), and we can't get hold of it for ages because the group reckon the time isn't right to mix any more!'

Whatever its background, the incense was just right for the ki-lin. The charcoal discs fitted neatly into the hollow back and stayed alight, while the incense curled langorously out of the mouth until the charcoal collapsed into a tiny heap of perfumed ashes.

Later, the 'group' dispersed and Don was hard put to get a new supply of the incense, or at least something like it, but he did find other things like myrrh to add to the list. Sadly, the Astrological Emporium itself is long gone, but I still have several precious jars of the incense. On very special occasions, such as Christmas and New Year, they are brought out. I sprinkle minute amounts of their contents on the charcoal inside the ki-lin, and the tangy perfume mixes with the smell of the fir tree and the freshly made clove oranges, and wood smoke.

At New Year, I use 'Sun', in the hope that its namesake will shine a bit through the coming months and, with deference to the Bwbach, a touch of 'Elvin'. As Mrs P says, 'You can't be too careful with him!'

That particular New Year's Eve, we sprinkled on a bit of everything in the incense larder and nearly asphyxiated ourselves, because by then there were faint glimmerings that the pilot programme had been well received by the new Controller of Radio 4, Monica Simms. So Mrs P and I were pulling out all our assorted tricks from the bag and wishing on everything in sight and spinning round like tops at the New Moon and spitting when we saw a single magpie and babbling happily when we saw 'two for joy' and, thus occupied, felt we'd done everything to help her make up her mind.

The liberal sprinklings of 'Sun' didn't keep the snow at bay that January however, and it came and went, sometimes quietly and sometimes spitefully in big drifts. It completely buried the new car in my life, a bright green Beetle to replace the old blue one with the embarrassing number plate FFU, which at least helped to find it a buyer.

The green Beetle wasn't really new of course, just less old than FFU. In fact it was already eight years old when I bought it, but at least it went, which FFU had shown an increasing reluctance to do. Its number plate was boringly sedate, although, in my overly superstitious mood, I saw significance in that too.

'272K,' I said to Mrs P. 'That's almost the STD code for Bristol, which is 0272!'

'So?' she said.

'Well . . . it might mean it'll be going to Bristol a lot.' I said, sheepishly.

'She's gone completely mad,' Mrs P remarked to Winston. 'You realize don't you my dear, that this is the Chinese year of the GOAT. Not a reliable animal at the best of times!'

So, when Michael Bowen rang that February day, when for once the frozen ground was melting and allowing the earth to breathe and send out a rich, earthy smell, headier than any incense, I was suddenly beset with a million doubts as he told me in his calm, matter of fact voice, that Monica Simms had commissioned a series of programmes to be called *A Small Country Living*.

I was right to be wary, because the next thing he told me was that he was retiring early and wouldn't be there to see the programme through and, when I drove to Bristol a fortnight later to meet the new producer, the bright green Beetle broke down half way there.

=Chapter 11=

For the umpteenth time the torch fell over and bleared its light uselessly into the straw. The ewe shot forward, causing me to lose my grip on the tiny hoof protruding from her back-end. The hoof slipped back into inner darkness and I lunged after the ewe before she careered off to the end of the lambing pen.

Edna's lamb was being born with one leg twisted back and I needed to pull the leg which was coming normally out past the elbow joint so that I could straighten the shoulder to give me space to grope for the other leg and correct its position. That done, and the second leg likewise extended, the head of the lamb would be able to follow more easily and if, as I suspected, it was a large head, I would have more room to help it out. Edna had other ideas about this, however, and for the past ten minutes, we'd been debating energetically up and down the big hayshed. The other black sheep watched our efforts casually, only making sure their lambs were tucked up safely out of our path. If they remembered their own disputes with me over the past week, they showed no sign of it.

I can never understand how sheep, bang in the middle of labour, can shift so fast. They've got it all their own way of course. In view of their supposedly delicate condition, one tries to hunt them down gently and dispense with flying tackles and brute force. So I had trotted around behind Edna, trying to persuade her to come back into the lambing pen. She'd escaped originally due to a couple of older lambs sneaking in through a gap and, while I was untying the hurdle to get them out again, she'd leapt up, pushed me over and gone free, in spite of the tiny hoof sticking out of her behind.

Finally, she'd paused in her hysterical gallop and I'd edged her back to the pen, but she continued to kick and wriggle, so that every time I got a grip on the little hoof, the torch went flying again. Not that I really needed the torch; the job I had to do was all done by feel anyway. I closed my eyes, gritted my teeth, told Edna she was a silly old bag and I was a worse one, got her firmly on her side and took a strong grip on the slippery hoof. This time I didn't mess about, but pulled it straight out till that blessed moment when the joints clicked past the elbow and the shoulder, and the full leg was extended. Carefully I felt inside for the other leg, hooked my finger round it, straightened it, and then pulled it out like the first one. There was now only the head to ease forward and, to my relief, it began to slide out sweetly of its own accord. At once, it seemed the whole body was there writhing on the straw and I placed it by Edna's head. Immediately, she began to lick the lamb, now taking its first gulps of air, and the blood drained back into my veins.

I shall never get used to that moment of relief when the lamb or kid is safely out, invariably convinced as I am, that something will go dreadfully wrong.

Edna paused in her frenzied ministrations to her lamb and heaved again. A second lamb slid out effortlessly into the world and, as Edna attended to it, I picked up the first one, anointed its navel cord with iodine, planted it near a teat and made sure it was suckling. Edna, in her new found glory, licked my hand, which was still covered with birth fluid, including me in her joy instead of treating me like the enemy. When both lambs were dry and fed and reeking of iodine, I picked up the redundant torch and went to get Edna a warm drink, some fresh ivy leaves, dried pennyroyal and a clean swatch of hay. She fell on them eagerly, blowing bubbles as she drank and tried to talk to her lambs at the same time, then devouring the food, for she'd refused either to eat or drink all day.

It was a harmonious scene in the lambing shed, now that Edna's dramatics were over. The straw was thick and golden and islands of black sheep, lambs tucked up behind them or slumbering on their backs, lay dotted in the hollows. At one end, however, sat a group of white sheep, their centre a huge, fat ewe, her own lambs nestled against her and, disposed around her, three younger sheep, one of them with another pair of lambs.

It was Minta and her family, creating a no-go area which any of the other sheep breached at their peril. Minta had been given to me by Myrddin, when she was a tiny pink and white scrap of a lamb herself. She had spent her babyhood with Dolores and Nana, when they were kids, and had grown up into a massive, greedy ewe with a domineering personality. Her daughter, Rosie, and she had both given birth to a female lamb the year before, and now the pair of them had produced twins, so that Minta's own personal flock was likely one day to overtake the Black Welsh Mountain one I was trying to build up. And there were times when I suspected she was creating her own private army to that effect for, unlike the other sheep who seldom retained a great family identity once their current lambs were weaned, Minta kept all her offspring and their children firmly under control. In fact, that year, when Rosie had been having trouble lambing, it had been Minta who'd stood bellowing for me to come and help and stood by anxiously till the lambs were safely delivered.

Minta's family in its turn, treated her with infinite respect, stayed close by her at all times and stood back to allow her the best of the food. At night, in the shed, they vigorously batted off any attempts by the other ewes to take Minta's chosen spot and then placed themselves on guard around her. Her current lambs were treated like royalty and even Rosie's new pair took their place uncomplainingly in a less favoured position. During the day, when the flock went out onto the field, one of her family would baby-sit all the lambs and I never had any fears about marauding crows on their behalf, unlike some of the others, whose mothers would leave them quite alone

111

while they wandered, absent-mindedly plucking at the grass, further and further away.

I could have wished to have found so much altruism in the goat-shed, but there, it was every goat for herself and any kid who overstepped the boundaries got its ears bitten or found itself being swung around by the tail.

Loveday and Little-Nana kidded early, had a female kid apiece and changed from silly young things themselves, into dignified matrons with responsibilities to bear. For the moment, they were producing enough milk for the kids and me to share, but later I would take the kids away from them late at night, milk their mothers in the morning, and then let them have the milk for the rest of the day.

Still hanging fire, however, were Whacky, Dolores and Minnie. Minnie I wasn't worried about, because she wasn't due for another three weeks, having been served so much later. Dolores and Whacky, however, were already over their time by quite a few days and although it was obvious that Dolores would pop at any moment, Whacky was still dancing about blithely unconcerned.

One night, when the activity in the lambing shed had been at its most hectic, it became obvious that Dolores would kid that night. Her udder was enormous and the pelvic ligaments, that run diagonally from the spine out to the pin-bones, were soft and soggy. I left her alone in the big stable and checked on her all during the night, but by dawn she still showed no sign of going into labour. Dizzy with sorting out lambs and ewes and still anxious about one or two of them, I left her for no more than half an hour, only to find, when I came back, that she had flung herself down on a sharp cobble stone, which she'd unearthed as she pawed at the deep straw, and dislocated her hip. Once again, it was a job for Bertie.

'I think there's one more inside,' he said an hour later as we contemplated the two kids he'd had to deliver with infinite care while Dolores lay exhausted.

'There can't be!' I cried. 'Those two are enormous. There isn't room for another one!'

'Well, there is,' he said, as he groped carefully inside Dolores' womb. 'It'll never win the shot-put at the Olympics, but it's there all right,' and gently he pulled it free, so tiny that it almost fitted into the palm of his hand. It gave a feeble little choke and a cry as he handed it to me to dry, for Dolores was too tired to bother with any more kids.

'It's GORGEOUS!' I breathed as the minute creature emerged from the birth slime into the most delicate mauve and gold colours.

And Gorgeous she's remained to this day, the fattest goat on the place, the greatest nuisance and the one photographers love best because she'll make silly faces for them. It was Gorgeous who invented post-natal depression in goats (or at least was the first to have it in public), got herself into *The Times* Colour Supplement and featured in the *Daily Telegraph* under the heading 'A Small Country Living With a Goat Called Gorgeous'. She's also had an amazing psychic experience, can convince herself she's in kid, just by the smell of billy goat, and produce a fantastic cloudburst exactly five months later, instead of milking through as she was supposed to have done.

'I'd shoot her if she was mine,' said Christine Palmer once, but, if I often agree with her, I just have to look at that beautiful golden coat, underpinned with lilac, or see her ridiculous face wrinkling its nose at me, and I forgive her much.

Her mother never liked her though. Dolores recovered, but gave all her love and attention to the other kids. Gorgeous, she gave it to be understood, got on her nerves. Far too small for a start and that colour was ridiculous!

But 'ridiculous' was a word which applied far more to the next arrival, still biding its time inside Whacky.

I have told elsewhere of my first anguished encounters with Whacky, who was still playing her double game of being the most delightful of goats by herself, but an absolute little swine amongst the other goats. Since I'd sold her mother, Nana, she had fought her battles alone and unaided, determined to be the odd one out and although her sister, Little-Nana, sometimes took her part, Whacky thanked her not at all. She continued at all times to defy Dolores as head of the herd and took every opportunity to drive the rest of the goats into wild disorder. As for me, she waited till I could see her very clearly before committing some heinous crime, like calmly shredding a hay net or chewing the wood off the main door, knowing that there was some fearful bond between us, so that whereas with the others, when they misbehaved, I merely got cross, with her I would lose my temper. Only then, her object achieved, would she simper sweetly and do exactly as she was told. It was all part of a Whacky plot to goad me into making a fool of myself. By the same token, if she was getting the worst of some fight with the other goats, which doubtless she'd started anyway, I would fly to her defence only for her to turn a face of total innocence upon me as if enquiring what I was making all the fuss about.

She adopted exactly the same attitude to the business of whether or not she was in kid. The fact that she hadn't come back into season for the past five months, was neither here nor there; goats are just as capable of having a false pregnancy as any other animal, as I've since found out with Gorgeous. Given Whacky's ability, conscious or unconscious, to cause the maximum of trouble, it was on the cards that she was certainly kidding in one sense if not the other. Whereas the others had bulges like badly packed overnight bags on either side, Whacky had a firm, uniform stoutness which might just have been plain fat, and the movements I'd felt in her side, too much wind.

'If she *has* got a kid in there,' I complained to Mrs P as we examined her for the thousandth time, 'it must be a midget. Whatever it is, it's not too active.'

'You may be very surprised,' replied my mother, who likes to bet on three-legged horses and lame dogs. Funnily enough, they often win for her too.

'Oh, I'll be surprised all right,' I said bitterly, 'but only if she produces anything at all, the little brute.' Whacky turned a knowing eye on me and sauntered off with a contemptuous flick of her tail.

It was a good week after she was due to kid, if at all, when Whacky showed signs of slowing up, as if she was carrying something else besides her own bulk. I felt her pelvic ligaments and they had softened enough for me to park her alone in the stable in which Whacky and I had fought our first battle royal when she was a kid herself and where she had lived so long on her own. She eyed the manger thoughtfully, decided that her new heaviness was too great to allow her to leap up into it as she used to do, and then she settled herself against the bales of straw lining the rough stone wall.

When I checked on her again, Whacky was still sitting in exactly the same position, calmly chewing her cud. She hadn't a care in the world and neither had I, till I noticed two small feet poking in and out of her behind.

'I think,' said Mrs P from the doorway, 'that that goat is about to give birth.'

I glared at her. 'Perhaps you'd better tell Whacky that then,' I said. 'She's not even having contractions!' And still the little feet waved in the air and still Whacky sat there chewing happily.

At last I intervened and took a gentle pull on the feet. Whacky looked round, yawned and concentrated on bringing up another wad of cud.

'Push! You rotten goat!' I yelled at her as the nose of the kid began to appear.

'Good grief!' said Mrs P in awe, 'she's not even bothered!'

'No, but I am. This is no tiny kid. It's enormous! Considering she's never kidded before, she ought to be feeling *something*!'

'Perhaps,' said my mother who, as a midwife for years, had seen just about every variation on the theme of birth, 'it's one of those cases where the woman feels nothing and the husband gets all the birth pangs. Only there isn't any husband . . . so who's having them?'

Whoever it was, it certainly wasn't Whacky, who stayed quite still and continued to cud happily while her child and I fought to get it born. Even when the huge shoulders came out, Whacky did little more than hiccup politely, and it was left to Mrs P and I to panic as the body of the kid continued and continued to come out. It was not only big-boned, it was almost as long as Whacky herself and it was like pulling, hand over hand, on a rope to deliver it. When it eventually arrived, complete, Mrs P and I could do little more than stare for a moment. As I freed it of the mucus and mess around its nostrils, it gave, not the usual petulant little cry of the new-born kid, but a deep, bass sound of complaint.

Whacky looked casually over her shoulder at the lump in the straw and seemed at last to realize that something new had happened in her life. She got up as I brought the kid round to her head and gazed at it thoughtfully for a second. Then, and then only, did some kind of maternal instinct rouse itself in her breast, for she started to lick the kid, slowly at first and then with increasing relish, and all the while the kid kept up its sonorous comment on life.

Mrs P and I collapsed onto a bale of straw to watch the little drama of mother and child unfold, until at last the kid began to struggle to its feet and go in search of a teat.

It sometimes happens that a goat, kidding for the first time, will be horrified at the idea of her kid suckling and it needs a fair bit of persuasion to get them to accept this final act of motherhood. I'd had a bit of trouble with the others already, so I was quite prepared for Whacky to reject the kid's attempt

'Damn,' I said to Mrs P as Whacky swung her udder well out of reach of the kid's questing mouth.

'Double Damn,' I said as she walked right away and made for the fresh hay I'd left on one of the bales of straw.

The kid, bowled over on its shaky legs by this abrupt desertion, gave another deep, unhappy cry. I got up to start the business of reuniting the two and persuading Whacky to stand still while her kid drank that precious and vital colostrum.

'Hang on,' warned Mrs P, 'she's coming back.'

'I don't believe this,' I whispered, as Whacky returned to her kid with her mouth full of a large swatch of hay. I'm glad my mother was there to see it, because I don't think anyone else would have believed it either.

Whacky paused, dumped the hay on the ground and began to nudge her kid carefully towards it.

'It's logical I suppose,' laughed Mrs P. 'As far as Whacky's concerned, she's feeding it the best she has!'

We managed between us in the end, the kid and I, to convince Whacky that kids drink milk and finally, no one made a better mother.

As for the kid, it was female and it was the tallest, longest and ugliest I've ever seen. She was also the most amiable, so that when I named her Ugly Bananas, it was all to do with her looks and her shape and not because she took after her mother in any way at all.

Chapter 12

All that April it went on snowing in fits and starts and the hay ran out. Down to my last couple of bales and word out everywhere that if you could buy it at all, it would cost a fortune, I couldn't believe it when an immaculate Volkswagen Camper lurched into the yard loaded down, inside and out, with big, golden bales of hay. Barely visible amidst all this amazing bounty, Andrew Naylor beamed and laughed and hooted his horn triumphantly.

Andrew and his wife Carry both worked for BBC television in Bristol. I'd met Carry a few years before when I was a guest on a television programme and I'd met up with her and her new husband again when I'd gone over to meet the producer of *A Small Country Living*, Sarah Pitt. Since then, they'd been to the farm a couple of times and had done all sorts of things for me, like mending locks and clipping the hedge and sorting out lots of practical jobs I'd made a mess of, in exchange for nothing more than the chance to wander about on the meadow. They'd been over the week-end before, when I'd been doling out wisps of hay to the horses and goats and wringing my hands in despair over how much longer it would last. A few days later Andrew drove into the yard with his surprise, grinning from ear to ear, and full of his story of how he'd taken time off from work to scour the countryside around Bristol to find me the precious hay and had even managed to get it at half the price I'd expected to pay. As soon as he'd unloaded it and had some tea, he was off again, the full hundred miles to Bristol, with the threat of more snow on the way, and got himself to the top of Mrs P's list of messenger angels despatched in answer to her efforts on the Hot Line.

It snowed again, even on May Day, and it was hard to believe that spring, let alone summer, would ever come back again. The best the weather could do was to go on blustering and raining and finally settle for fog once more. When that cleared at last, I saw that the swallows had come back and two of them were twittering at me impatiently to open the top half of the door leading into the long cow shed, which is where the goats now live. They had made their nest for the last couple of years, high up on the cross beams of that shed and, for as long as they are in residence, the top half of the door is never closed. The goats and I are used to being dive-bombed by swallows all summer and we're pretty good at ducking their offerings from on high.

The problem with leaving the door open, however, is that the pigeons begin their annual campaign to get back into the cow-shed to nest along the stone ledges lower down, and from thence into the main barn.

I am extremely fond of my pigeons. They're a wonderful mix of racers, tumblers, fan-tails, turbots and hollecroppers. All of these breeds had been tried by the Artist and me in our first years at the farm, but being less than careful about keeping them confined in their pairs, somehow all of them mated up with the wrong kind. Over the years, the mix of the various talents meant that the ultimate pigeons could fly very fast, tumble, fan their tails, pout till they nearly fell over backwards, and sometimes grow little crests at the back of their heads. Their colour is mostly white, with the odd cap of grey, or streaks of grey and fawn on wings and tail.

The pigeons fly to great heights all over the valley and the far distant specks of white, or the sight of them, wings up, gliding in to land on the barn roofs or tumbling giddily into the yard, are as much a part of the place as the big ash tree itself. If I'm out across the fields and see them above, I call to them for the fun of seeing them put on the brakes in mid-flight and do an aerial U-turn to come home and wait for me. No matter how often this happens, I get a thrill up and down my spine, not so much on behalf of the person I am now, but because of the lonely, frustrated child I was one blindingly hot summer, long ago in a small, dusty town in New South Wales.

Mrs P and I were living in the one and only hotel; a big ramshackle place with wide wooden verandahs, iron beds that creaked, cupboards that never shut properly and primitive bathrooms where the taps either didn't work at all or dripped relentlessly. It was during the time when my mother roamed all over the Outback taking nursing jobs. Sometimes I was left with relations in Sydney, sometimes at boarding school but most often, because I hated the other options, I tagged along behind her and went to whatever school was on offer. This particular time, however, it was the summer holidays and, all the local children being mostly back out on their farms and stations, there was little to do except hang around the gulley at the bottom of the long, dry yard of the hotel. And then the Rodeo came to town.

They came in from everywhere to see the show, which was held down on the sports ground under a big marquee, stifling hot even at night. They called it a Rodeo, but it was more of a travelling circus, with clowns and trick ponies and a magician with lots of white doves which he pulled out of all sorts of places and threw up in the air to flutter out of sight. It was the white doves I remembered and dreamt about all night.

Next morning at the hotel, there was the magician himself, sadly without his glittering cloak, tucking into a vast cooked breakfast just like everyone else.

'Did you like the show?' he asked me with his mouth full.

I nodded, eyes wide.

'She especially liked your doves,' said my mother. 'She's been talking of nothing else.'

The magician looked at me searchingly. 'Would you like some doves?' he said at last. 'I happen to have a couple spare. They don't want to learn their tricks like the others. Too interested in billing and cooing.'

Before he left, the magician gave me not only the two doves, but a stout wooden cage for them. He showed me how to water and feed them and assured me that if I kept them locked up for a while, I'd be able to let them out eventually and they'd fly back home to their cage. My friend the cook and I found a cool place in one of the outbuildings for the cage and everyone in the hotel came to admire my birds and give me advice.

My dream of white doves lasted just one day. Over-confident, or perhaps not then used to the sneaky way animals have of taking advantage of you, I opened the cage just a bit too much and, as one bird, they streaked out of the gap. I can still feel today the total despair and sense of loss as I watched those doves fly up against the sun and out of sight.

Everyone told me that all I had to do was put salt on their tails to catch them. Even my mother and the cook joined in the joke, till the latter took pity on me and helped me rig up their cage out on the yard, with a string attached to the opened door and a long line of corn leading to it. For days and days I sat there waiting for my doves to come home, but only the magpies came down for the corn. Word spread round for miles about the doves and it seemed that everyone had seen them . . . north, south, east, west. So I wandered, bag of corn in hand, calling and calling, and once, just once, I saw a flash of white wings, but it was only a flock of cockatoos.

The loss of those doves haunted me all my childhood and beyond, and for years I could never see white birds without remembering those hot days, sitting patiently on the yard with the string in my hand, or the endless wandering and hoping. Which is why the sight of a whole flock of them swinging home across the valley gives me such a kick.

Not that our relationship is all sweetness and light. When the Artist acquired the first pair of racers from Gerald Summers, he'd made a proper, wooden nest site for them in a cow-shed, of all places, and there they'd bred. From then on the cow-shed was home to each succeeding generation, and from there they'd spread into the main barn where their droppings fouled the feed bins and the hay and straw. When the Artist left, it took me a solid autumn and winter to

120

persuade the pigeons that there was a perfectly good, custom-built dovecot attached to the other end of the barn.

It's a stout lean-to, with a proper slate roof and a square hatch cut into the thick stone walls for the pigeons to use as an entrance, and I've often wondered how it came to be there. It's unlikely that there were any pigeon fanciers here before me, unless they kept them for the table. Whatever its origin, the pigeons took a long time to regard the dovecot as a suitable residence, and they grew more determined than ever that the main barn offered much more scope for their eternal breeding.

I must explain that the cow-shed, the main barn and the old stable, are all interconnected, so that a door left open in one, gives access to the others. When it came to evicting the pigeons, therefore, they had all the advantage. I spent a lot of time teetering around on bales of straw trying to dislodge them from the beams high overhead, with the aid of a long pole. Dislodged, my feathered friends would lazily glide to another beam or, even worse, through the hatch in the wall of the stable. By the time I'd got round into the stable, they'd have sailed happily back to their original roost. I wore them down eventually, but if the doors of cow-shed, barn or stable were left open for a moment, back they'd come and it was all to do again.

When the swallows decided, after having deserted the farm for a few years, to make their nests once again in the cow-shed, I had to decide carefully between swallows and pigeons, or no pigeons in the barn and no swallows either. So every summer I grit my teeth, get out the long pole and do battle with those pigeons still carrying the inherited memory of blissful matings in the barn.

For one pigeon however, no amount of keeping the doors shut, or wandering about with my pole, did any good. He could find his way in through the smallest crack and managed to teach a few of the others to do it too. Blatantly they queued on the ground outside the barn, each slipping in through a gap in the bottom of the door, one by one.

This particular pigeon didn't even have the excuse of race memory. Old Toughie was lucky to be alive at all, after being knocked down by a Number 9 bus in Kensington Church Street. He was one of the hordes of pigeons who inhabit the Gardens and sometimes come out to brave the traffic and search for crumbs around the best patisserie in London. I'd found him, wing broken and covered in blood, as I was emerging from one of my illicit trips to that patisserie in search of a bag of their wonderful Chelsea buns. Diving under the wheels of another Number 9 bus, following close on the one which had swiped Old Toughie off his meal, I'd rescued him and taken him back to my flat and patched him up. From there of course he'd come to Wales and the safety of the country and he'd stayed to make me regret it every day.

He was an ugly, dull, grey, in colour, with not a single item of personal beauty about him. In spite of that, he caused untold marital dramas by trying to elope with every female pigeon on the place. Pigeons are usually the most faithful of birds, but for some reason, those females were quite prepared to give him what he wanted. Perhaps it was something about his street-wise ways which could get them back into the barn, their palace of delights.

For me, Old Toughie had long since ceased to have any charms at all. I tried discreetly to remove him from my life by taking him with me whenever I went away, releasing him far from home. He was always there when I got back, trying to figure out a way of getting past my latest bit of carpentry on the barn door. I longed for another Number 9 to appear magically and finish the job of squashing him. I was still soft enough to baulk at doing it myself.

'There's only one thing left,' I told Mrs P. 'He'll have to go back to London. Maybe the Gardens and the patisserie will keep him occupied. The problem is, getting him there!' For I'd long since given up even a remote connection with London myself.

'What about the lady who's coming to collect Little-Nana?' she suggested. 'She might take him back for you.'

'What a very good idea!' I said, and reflected that getting rid of Toughie might be a slight compensation for the loss of Little-Nana.

There comes a time, over and over again, in every goat-keeper's life, when decisions about which goats to keep and which to sell, have to be made. For the more commercially minded, it all comes down to milk production and age, but for people like me, who get overly fond of their goats, it's far more complicated and, if it wasn't for lack of space or the sheer cost of feeding them, most of us would be overrun with goats. Actually, most of us are anyway.

I'd been vaguely thinking that the goat shed would soon be getting a bit overcrowded with all the new arrivals; added to which, the flow of milk was beginning to swamp us, even sharing it with the kids, the dogs and the poultry. Selling the milk was something I didn't really feel able to do. For a start, there wasn't the demand for it locally, but, most important of all, I felt that to do it properly would mean getting a special dairy-complex with sterilized equipment. I was appalled sometimes when I saw the haphazard way some goat-keepers treated the production of milk for sale and it always amazed me that they hadn't poisoned more people in the past. Now, just a few years later, goats and their milk are being taken more seriously and things have improved a lot, but personally I've never had the time nor the equipment to take the risk of selling our milk.

Time in particular was something I was going to be very short of if I was to go back into broadcasting regularly. I'd already decided that Doli's daughter, Gwylan, would have to be sold to some friends who were anxious to buy her, and it seemed reasonable to cut the goat numbers down as well.

I hadn't given the matter too much thought however, still too bereft at parting from Gwylan, even though she was only going a few miles away, when Caroline rang. She was a journalist who lived with her family at Greenwich, where they had the use of a very big garden. They had decided to use part of that garden to keep a goat and, as Caroline was staying in the area for a few days, she wondered if I knew of one she could buy. It would have, she assured me, the best care and attention, never be left on its own and have plenty of good grazing.

I'm never very keen to sell a goat where I know it will be the only one on the place, but Little-Nana was rather like her sister Whacky, in that she much preferred people to goats and, given the choice, I'd rather sell a goat into a life of solitary luxury than a communal one of starvation and misery, any day. There was, of course, the problem of Annie, her kid, which Caroline didn't want, but I was already bottle feeding Gorgeous, so she could join her.

When Caroline and her children met Little-Nana, it was a true meeting of minds, Little-Nana turning her back on me and devoting herself exclusively to charming them to death. She hated it when I took her back to the goat dorm.

Caroline came a few times for milking lessons and a crash course on goat management and finally arrived to collect Little-Nana for the long trip back to Greenwich. It was not very easy packing a large goat into an old Morris Traveller with two children, two adults and a mass of luggage, but Little-Nana seemed, apart from an initial nervousness, to be perfectly happy peering at me out of the back window.

'Could you also fit in a small box?' I asked Caroline, 'and would you mind opening it somewhere in the middle of London?'

'Not a bomb is it?' she said.

'No, just a bus-proof pigeon,' I replied and went on to tell her the whole sorry tale of Old Toughie.

'Sounds as if he really does need the wider scope of London,' said Caroline. 'Right! Where is this macho individual? I'll let him out in Greenwich Park and he can do what he likes from then on.'

She rang a few days later to say that the journey home had been only a minor sort of nightmare, that her husband's carpentry skill was now in serious doubt due to the way his 'goat shed' had collapsed under the impact of Little-Nana, that she, Little-Nana, had partially devoured the manuscript he was working on, but was otherwise thriving, happy and well. Toughie had been released as promised and had last been seen heading in a vaguely westerly direction.

It took him eight months to get back to Wales. Only, by then, I'd had the barn door mended properly and there was a new and thrusting generation of pigeons to overtake him in the mating game and, in truth, his adventures had wearied him for the chase somewhat. He spent his days in peaceful retirement with an aged lady which none of the others fancied and their devotion was wonderful to behold.

Long before Old Toughie's less than triumphant return however, I had a large and lonely horse on my hands.

Doli is not the most demonstrative of animals, unless there's food in the offing, or the time is right for the stallion. About the only time she gives any indication that she does have a heart tucked away in her vast bulk is when her foals are still young. Until they are weaned, she is, if anything, over-conscientious with them and pampers them shockingly. If they expect the same treatment when they return to her after weaning, they're in for a nasty surprise.

124

Doli bites and bullies them into line and the slightest insubordination has her massive hind-legs flying out like pile-drivers, right on target. She also steals their food.

By the time Gwylan was two years old therefore, I thought that the grief of parting with her would be mine alone and indeed, to look at Doli, she seemed fairly indifferent to the absence of her daughter. She pottered around with Blossom at her heels and seemed much the same as usual, except that a few days later she had a slight dose of colic.

'Probably worms,' said Bertie after he'd given her an injection to remove the blockage.

And then Doli went lame. She only went lame when I was looking at her, so that when I called Bertie in once again, he looked at me severely and said:

'There are such things as animal hypochondriacs you know.'

'Who? Doli?'

'No, you!' he replied. 'Anyway there's nothing wrong with her feet or legs that I can see. Trot her up and down again!' I stumbled a bit, but Doli trotted straight and true.

She was off her food and dead lame again the next day when Paul, a policeman with a keen interest in horses and a few of his own, came to see me unexpectedly and caught her in the act.

His verdict was pretty much the same as Bertie's, except that he could see I wasn't completely imagining it.

'She could be bored and lonely without her daughter,' he said. 'Have you ever thought of mating her to a Shire? I know a beauty called Wishful Wanderer down on the Gower.'

'Will he have to cover her in hand?' I asked.

'Damn yes! Can you imagine a great Shire running free with his mares?'

'Well, he'll be Wishful Thinking if he's got any hopes of getting Doli in foal I fear. She never seems to hold when she's served in hand.'

'Why don't you give it a try anyway,' said Paul. 'She's only going to hang about here being colicky and lame for something to do. If you like, I'll have her to stay with my horses and take her over to Wishful Wanderer for you a couple of times.'

And so it was arranged for Doli to stay with Paul for the summer, to try her luck with Wishful Wanderer and to leave me free of worrying about her while I finally settled my mind to getting the first series of *A Small Country Living* on the air.

=Chapter 13=

It's hard to do a sensible interview when a llama suddenly thrusts its supercilious face up against yours and breathes heavily down your nostrils. It's difficult to go on talking calmly about the practical aspects of breeding Welsh Harlequin ducks, when two of them are mating energetically in front of you; it's not easy to keep your microphone on an even keel when an enormous Toulouse gander is chasing you, great wings flapping and hissing fit to burst. And you don't even try to stop screeching with laughter when a Golden Guernsey goat reaches over and snatches the piece of paper, from which her owner has been reading a classical quote about such golden goats, and munches it up loudly. Bless 'em, the animals all performed right on cue and, if I'd had any ideas about the programme being a deadly serious one on how to make a small living in the country, I forgot it after my very first trip.

Neither do I worry too much any more about where all the stories

...me from; the best ones find me somehow. Those that I
...spend hours on usually turn out to be duds. And so, I
..., do many of the ones I follow up from letters (unless
...n written by a disinterested party about someone else),
for n... I've found the truth of those lines by W. B. Yeats 'The best
lack all conviction, while the worst are full of passionate intensity',
it's since I've been doing *A Small Country Living*. It's the people
most surprised that I'd be interested in them at all who have
produced those special moments every broadcaster dreams of; it's
been the casual, thrown-away remark which has led to the piece
that's filled the mail-bags to overflowing and been remembered,
years afterwards.

It hasn't been easy either, to convince the two producers I've
worked with that the big stories are better covered elsewhere (and
usually have been), but when I put 'small' in the title, I meant it in
every way. It may not seem of world-shattering importance to
discuss value for money in wellington boots, or the best way to get
your whitewash to stay on the buildings, or how to keep your septic
tank in good order and odour, or even what to do if a bee gets in your
ear or your ducks go down with the July Sprawls, but a surprising
number of people found it riveting.

There's also a lot of sex. My producers, who don't usually hear the
contents of the programme till I appear with it all banded up and
ready for linking, have got used to keeping their faces quite straight
when rams and bulls, goats and horses, ducks and geese and
chickens, all uninhibitedly make their passionate presence felt as
their owners and I calmly discuss their other qualities. The animals,
as I've said, have a wonderful capacity for chiming in on cue except,
that is, for the ghost pig which used to haunt the author BB's
lovely old toll-house in Northamptonshire, so we had to leave it
to the listeners' imagination with a bit of help from the sound
library.

Not that the animals ever quite came up to the outrageous stories
Fred Archer told. I'd ask Fred to take part in the programme before I
actually knew how it would turn out.

'I want you to be a bit of light relief,' I'd told him. 'Although every
interview must also have some information in it, if only, for example,
what village life was really like, pongs and all!'

128

Fred's memories of village life and farming in the Vale of Evesham went back to the twenties and what he hadn't actually seen, he'd heard about and stored up in his memory. He had so much in there, by the time he was in his forties, that he'd come in at the end of a long day on his farm and written it down in books which were successful enough for him eventually to become a full-time author. With his rich country accent and brilliant ability to recreate a scene vividly, he was a gift to any broadcaster, and I was lucky enough to know him well after we'd done several feature programmes back in the days when I was working in London. He was one of the first people I thought of when the new programme was in the offing, but even I hadn't realized how many incredible stories that memory of Fred's had tucked away. If Sarah Pitt, my original producer, got used to the uninhibited behaviour of the animals I encountered, she was still holding her breath about what Fred would come up with, when she left four years later. The listeners, however, loved his voice and his stories and, if that first programme was a success, Fred had much to do with it.

For a success it was, in spite of being tucked away on a mid-afternoon slot and quite unpublicized. The morning after it was broadcast, there came a pile of letters which people must have written the moment we were off the air and then run like the wind to catch the afternoon post. All of them were ecstatic.

When Sarah rang to tell me, her voice quivering with excitement and saying things like, 'I think we've got a hit!' the only person who was quite unimpressed with Mrs P.

'I never doubted it,' she said blandly. 'The way you've been wailing and worrying was a complete waste of time. You simply have no faith.'

'Yes mother,' I said and went back to the old editing machine to get another programme sorted out and spend the next week wailing and worrying over it.

By the time autumn came, I was a nervous wreck and, with the series at last, feeling not at all like catching up on all the farm work I'd neglected. Apart from everything else, there'd been a black puppy, with small shark's teeth and strawberry pink pads on his feet, to keep me awake at night and chew his way through the furniture by day. He was Merlin's great nephew and his birth had been a saga all on its own.

In the months following Merlin's death I had cursed myself over and over again for not breeding from him. Somehow there had always seemed to be plenty of time. He was one of those rare dogs who had never been constantly on the look-out for a likely bitch, and on the one occasion when I'd been asked to let him mate a charming little lurcher, he'd been far more interested in what her owners and I were having for tea than in getting on with the job. Eventually another dog had taken his place and, to be honest, I wasn't really sorry at the time, because once a dog is used for stud, it usually puts a repeat performance at the top of its list of priorities. I had intended, though, to find a whippet bitch to mate him with eventually, if only to keep his genes going. One way and another, I just hadn't got round to it, so that when he died, I didn't even have the consolation of at least owning one of his descendants. Desperate, I began to ring around to see if, by some chance, one of Merlin's relations had bred.

The first person I got in touch with was his breeder, but she was equally unhappy because his father Pluto, from whom he'd got his looks and character, had been killed recently. She had given up breeding whippets some time ago and had none of his family left. She did give me, however, a list of people to whom she'd sold Pluto's pups over the years, but their stories were equally sad. Several of them did say that if I ever found any of Pluto's descendants to breed from, they'd be very interested in having a pup. At last I advertised in *The Whippet*, a small magazine put out by the Whippet Club. There was one reply from a lady who lived not far from where I'd found Merlin himself and who had his sister and her two daughters. Ruth Overall would be delighted to mate one of these daughters to a stud dog of our mutual choice. We chose a black dog call Koh-i-Noor, a winner on the track and in the show-ring. Fourteen months after Merlin's death, the litter was born.

'The only pup who looks like your Merlin is black,' Ruth told me, 'but he's got exactly the same white markings and is certainly the best and chattiest of the lot. There is a little blue bitch. Are you sure you wouldn't like two?'

'I'd love two,' I said, 'but Mrs P is threatening to go on strike.'

'Well at least all the pups will be sold thanks to those people you got in touch with. Now, I've organized a prefix especially for them and I'd like to register them myself, so what are you going to call yours?'

'Wild Wizard,' I said.

He arrived, brought from Hertfordshire with one of his sisters, in a huge box marked Roverall Wild Wizard. His sister, Blue Pie, had been dropped off and met at the Kestrel Inn on the A40 and, although I've never seen her, she too was to make her contribution to our lives a few years later.

When I opened the box the small black pup crossed his eyes at me and honked. Apart from the fact that he was black, he was Merlin all over again. I tried calling him Wizard for a few days, but found myself reverting to Merlin and Merlyn he has stayed, except that I spell it with a 'y'.

He looked like Merlin, he was highly intelligent, he could make a faint imitation of the range of sounds his great-uncle excelled in, but he was not the same dog. Where one had been all light, this one was the dark side of the moon and his will and mine seldom coincided.

Neither did he take kindly to the idea that Winston was head of the pack. He'd obviously been the boss of the other pups and Winston, less than the size of those pups, was his first target. Winston responded furiously and the two have had an uneasy truce ever since.

Busy as I was with the programme all that summer, I snatched moments to take Merlyn out on his own to teach him to stay at heel and he learnt to retrieve by torchlight. I threw the ball down the beam of the torch, and ever since, the sight of the torch excites him. In fact all his training was done either by moonlight or torchlight and I would see him running across the top of the field as a dim shadow or standing silhouetted against the pale translucence of the summer night.

In the house, it was like owning two separate packs. There was Winston and his adored ladies, Bea and her daughters, Misty and Gloucester. Her son Boy had gone to live with the Naylors in Bristol. He'd never really been happy since his hero Merlin had died and, when Carry and Andrew began to come regularly to the farm, Boy decided they were what he'd been looking for all his life. At last they asked if they could have him and he hopped into the VW Camper without a backward glance. He doesn't even like coming back on visits in case they leave him behind. So that left just the three bitches and Winston on the one side and Merlyn, determined to be the only dog in my life, on the other. Even today, when the pack has undergone more changes, he is still the odd dog out. Perhaps the fault was mine.

'You can never hope to replace one dog with another,' Imogen Summers had warned me. 'It's not fair on the new dog to keep comparing it with the other one. Your other Merlin was a "one-off". Treat this one as something quite different. One day, if you're lucky, you'll find another really special dog, but don't bank on it.'

But I didn't listen. This one looked like Merlin; he was the same family; he must come up to the same standards. Poor little dog, he never has. Instead he's cost me more in anguish and vet's bills than the rest of the dogs put together.

He learnt very early on that the slightest cough from any of the dogs or cats would send me into overdrive. Ever since Merlin's death, the fear of bronchitis turning into pneumonia haunted me terribly, and a cough brought out hot water bottles, blankets, panic and pampering. So, if he thought life was getting a bit boring, or Winston the upper hand, Merlyn coughed. If he was lucky, it might mean a trip in the car to see Bertie and have his lungs listened to and be made a great fuss of. And certainly for a while it meant extra rations and sleeping up in my room on the armchair with cushions and blankets.

He is also accident prone. When he was barely six months old, he rushed straight through a bundle of ancient barbed wire lurking in the depths of the wood. For years I'd been hunting the stuff out and getting rid of it, but this lot had stayed hidden in the undergrowth till Merlyn encountered it on his way after a rabbit which was making a hasty exit into the forestry.

Merlyn disappeared into the wood an incredibly beautiful composition of black shiny fur and taut rippling muscles and he came out of it a bloody mess. Not that he'd felt a thing. When I called him he came to me tail wagging and prancing joyfully.

I froze when I saw the tattered skin hanging in strips and saw the jugular vein in his neck pumping, intact but horribly exposed. We were a long way from the house and there was no way I could pick him up without adding to that terrible damage.

'Good boy,' I quavered. 'Home! Run home!' And run we did,

132

Merlyn treating it all like some wonderful game, till we were within earshot of the cottage and I started screeching at Mrs P to get a clean sheet ready.

'Don't look,' I told her as I gently bundled the sheet round Merlyn and tried to keep him still. 'Ring Bertie fast!'

'I think,' said Bertie, a frightening couple of hours afterwards, 'that this bit belongs here. What do you think?'

We were sitting on the sofa with a heavily sedated Merlyn lying between us on a pile of sheets and towels. Both of us were trying to work out a jig-saw of strips of skin.

'I've never had any ambition to make a patchwork quilt,' said Bertie sourly as he stitched, 'much less a patchwork dog! As for this cut here, how it didn't rip his jugular to pieces is a miracle. Right! That's the last piece. I think he looks very fetching. Like Frankenstein. He's lucky to be alive anyway, and so are you by the look of you. Do you know you've gone a really fascinating shade of bile green!'

Merlyn healed beautifully and managed to commandeer the armchair in my room permanently in the process. Another sadly damaged and equally disruptive animal left, however, to fulfill a heavy schedule of appointments elsewhere.

Not that Gandhi was disruptive in himself. He was a very kindly, affectionate old lad. Big, mind you. So big that when he stood up on his hind legs , he could lean his elbows on the tin roof of the pig-sty and gaze benignly across the valley, ignoring the frightful, wanton display going on below the wall. I was glad, nonetheless, that Graham Young had made that wall higher and goat-proof. I doubt if Gandhi himself would have bothered to try and clear it, but the smell of him turned every single one of my goats, young and not-so-young, into raging nymphomaniacs.

They'd been sitting around, calmly decorating the view, basking in the glory of an autumn sun, eyes lazily turned to the rich tangle of golden woods climbing the slopes, when Christine and David Palmer drove their trailer into the yard. One by one, ears shot up and heads swivelled. Long, aristocratic noses inhaled luxuriantly and, as one goat, they rose, mesmerized, and drifted towards the gate. It was Gorgeous who gave voice first, a long passionate, ululating shriek that echoed around the buildings and nearly deafened us. It was a sound I got to know well over the next few weeks.

'I should have warned you to lock them up!' cried Christine as she got out of the van. 'I'm afraid he whiffs a bit.'

'Whiffs is not exactly how I'd describe it,' I said, backing off. 'Pongs to high heaven, more like. I'm not used to your chaps smelling so much.'

It was true. Male goats in the rutting season do have a strong musky smell which most people find appalling and other goats enchanting. Apart from their own innate mating perfume, the stud goat adds to it by peeing all over his front end and rubbing his face and beard in it. The general effect is fairly awful and some goat-keepers, like Christine, minimize it by washing their male goats frequently and doing all they can to make them fairly acceptable to the humans at least. Conky, Chorister and Cavalier were, all things considered, far less objectionable than they'd have liked to be, but there was a pretty fearful aroma coming from the trailer which contained Gandhi.

'Yes, I'm sorry,' apologized Christine, 'but David and I have just collected him and I'm afraid we didn't get a chance to clean him up. Most likely the journey's upset him a bit too. He's probably been peeing like mad all over himself. If you wouldn't mind giving him a bit of a wash, he'll soon smell better.'

'You mean you want *me* to wash him?' I said aghast.

'Well, see how you go?' said Christine, looking at my ashen face. 'He's really very amiable considering what he's been through.'

For Gandhi had had a sorry life. Christine sold him as a kid, keeping his brother, Cavalier, herself. His pedigree was impeccable, she hadn't sold him cheap and, as far as she knew, Gandhi was destined for a life of honour and good living as a stud male. If she'd had any idea that he'd be left tethered alone out on the common, at the mercy of a mob of dogs which would savage him and almost tear his ear off, he would never have left her place alive. She had finally heard about Gandhi's fate on the goat-keeper's bush telegraph and dropped everything to go and buy him back. It was while she was negotiating ways and means of collecting him, that I'd rung with my annual query about getting the goats in kid.

'Why don't you have Gandhi for a couple of weeks?' she'd suggested. 'We have to pass your way when we've collected him, so

I could drop him off and then come back for him later. It will save you transporting your lot to me.'

And so it was arranged that Gandhi would occupy the Goat-Alcatraz until such time as my goats were well and truly served. Now he was here and many doubts suddenly assailed me. Apart from his smell, which was almost visible, Gandhi was enormous and his tattered ear added depth to his lascivious leer.

'I don't think,' I muttered fearfully, 'that I'll be able to control him.'

'No need to,' said Christine bracingly. 'Just hang his hay net over the gate and you can tie a string to his bucket and hang that over too.'

'What about the girls?' I said. 'I mean, how am I going to supervize this mating business?'

I'd already seen how reluctant male goats are to bid farewell to their little bit of wedded bliss and had visions of Gandhi flatly refusing to let me have my goats back, once he'd got them to himself.

'Look,' said Christine soothingly, 'I'm sure you'll manage. If things get desperate, I'll come and get him. Meanwhile, let me have a bucket of water and a few old sacks and I'll clean him up for you. Would you have a scrubbing brush I could use?'

While she washed and brushed him, Gandhi stood there blissfully and emerged, if not quite the elegant, well-groomed creature that his brother Cavalier was, at least a uniform creamy-fawn colour and certainly less offensive to eye and nose. He also totally ignored the deafening crescendos coming from the goat shed where Christine's husband, David, and I had finally managed to incarcerate a bunch of manic females, smitten with collective lust on a stupendous scale.

For the next three weeks, Gandhi was the least of my problems. He was a perfect gent, except for the time he took a sudden, intense fancy to a strong, self-styled he-man who'd come to do some fencing for me. But his intended mates were frightful. Animals, unlike humans, are usually very circumspect about sex, only indulging in it in its proper time and season and then waiting till the female is cycling, but for three whole weeks my goats made a proper exhibition of themselves. They congregated every morning under Gandhi's wall; they danced about like so many hideous Salomés in front of him; they called, they pleaded, they cajoled and they threatened. As each one came into season properly (and in their current mood it was hard to determine that exactly), I had no fears that Gandhi wouldn't let them go. He was only too pleased to get rid of them.

It was Whacky who finally did for him though. He was spark out on the floor of the pig-sty when I went to get her, and she was still biting his ears and yelling at him in the most frightful fish-wife voice, till I had to give her a sharp clip and drag her, screaming defiance, back to the rest of the goats.

I missed Gandhi when he went, but I think he was glad to go.

'He looks well,' said Christine when she came to get him, 'but a bit tired.'

Gandhi gazed at her blearily and I could read his expression very well. Quite simply it said, 'A mob of mad dogs I can handle, but females like this are killers!'

At least he'd done his job and all the goats he'd served were in kid. Doli, however, returned from her summer holiday and her brief encounter with Wishful Wanderer, fit and healthy, but definitely not in foal.

Chapter 14

I hate the wind. I hate it because it tears and rips and destroys; because it turns my little world into a house of cards; because it torments my poor old ash tree which roars and howls in pain; because it twitches the nerves with cold; because it drives the rain bone-deep so that nothing can keep you dry; because it turns the snow into a terror, or, on hot days, dries the earth to choking dust.

I don't know which particular wind, or combinations of winds, I hate most. There's the big, blundering one that comes from the south, bursting down the valley and railing at the back of the house, playing horrible defiant tunes on the Rayburn pipe, or bellowing down the stone chimneys, like the great bully it is. There's the east wind, with a cry like hags on broomsticks, screeching malevolently at the front windows and nagging and plucking. The north and the west leave me more or less alone, but I hear them roaring overhead to hit, like distant traffic, on the hills opposite. But if north or west team up with south or east, nothing can save the little cluster of buildings and the huddled animals from their triumphant howls and the flying slates and branches.

The only comfort I get, at times like this, is when the postman says:

'It's a lot quieter here than it is up on the top. You're a bit sheltered here you know.'

But the curious thing is, that if I *was* out on the top of the hills, I probably wouldn't mind the wind. I might even enjoy it and watch it swirling the clouds or tossing the birds and revel in its power. For my hatred of the wind stems from responsibility; the need to protect the stock from it; the urgent necessity of repairing the damage it does; of agonizing over what might happen if the ash tree, standing guard over house and barns, should finally give in to the terrible rage tearing at its branches. Such responsibility has robbed me of my old delight in walking in a thunderstorm; of running with the wind; of marvelling at the drifting beauty of the snow; of listening, content, to the sound of rain beating on the roof. In other people's houses, the wind itself is a romance; a creator of atmosphere, as it was in the old farmhouse near the village of Myddfai.

It was keening quietly, but persistently, around the windows, making a strange, otherworld background to the ponderous ticking of the big clock and the sharp crackle of logs on the fire. The old gentleman spoke in his deep voice which sometimes cracked into the high pitch of age, slowly and with long pauses which the wind and the clock and the fire filled with sound.

It was a neat room, with an old dresser full of china Toby jugs, a long oak table and curtains made of Welsh tweed. The space above the fire had been filled with the bright, clean stones of a new chimney, running up through the old one, and brass fire-irons shone on the hearth. It was, in spite of the antique furniture, the Toby jugs and the tweed curtains, quite a modern-looking room, but our conversation and the keening of the wind, took us back to the time we were seeking; the time of the Physicians of Myddfai. The farm itself is supposed to have belonged to one of those Physicians and the old gentleman assured me that his herb garden still existed by the house.

The little village of Myddfai is the setting for the other part of the legend which begins in the tall, old farmhouse, diagonally opposite to mine, where my friend Gwynneth lived, and up by the shores of Llyn-y-Fan-Fach, the Little Fan Lake.

The first part of the legend has its counterpart in other places. It is the story of the Lake King's daughter who marries the poor farmer's lad, bears him three sons and then returns to the lake when, quite inadvertently, he strikes her three times. But the Legend of Llyn-y-Fan has this other part to it. The Lake Princess returned to her three sons and taught them the healing virtues of various herbs growing on the mountain near Myddfai and, in time, they became famous for their cures, and physicians to the king himself.

Whether they were related to the Lake Princess or not, the Physicians were real historical people and their herbal remedies can be read in old manuscripts. I'd already been up to the National Library of Wales, in Aberystwyth, to root about amongst old documents and to pick the brains of Dafydd Ifans, the Assistant Keeper of Manuscripts. It was all ostensibly for a programme called *Legend in Landscape*. Actually it was just a glorious excuse to go bothering people in my own search for the origins of the legend which seems, somehow, to have a lingering contemporary presence in the valley, far more real than the big orange tractors or the grim, grey sheep-sheds which have swamped the little white farms.

Myrddin came with me to Aberystwyth, glad of a chance to indulge in his own hunt for the history of the parish, its old field names and the identity of the forlorn heaps of stones where once a farmhouse had stood. He was like a child in a sweet shop, ordering photocopies of old tithe maps and parish accounts, quite oblivious of the awful bill he was running up.

We came back, darting along obscure forestry roads as usual, both of us laden down with historical goodies and talking about our finds, each far more interested in their own research than the other's; Myrddin in his fields and long-forgotten little trades and crafts (including the mill which used to be on my own stretch of the river), and I in the finer details of the story of the Lady and her descendants.

Dafydd Ifans had given me a copy of the full legend as first written down, with the two parts of Lake Lady and Royal

Physicians fused into one story, by John Williams, known as Ap Ithel, in 1861.

He also gave me a copy of a fascinating essay written by Morfydd E. Owen entitled *Meddygon Myddfai: A Preliminary Survey of Some Medieval Medical Writings in Welsh*, an essay which poured scorn on the idea that the two stories were connected, but which nonetheless attested to the fact that the village of Myddfai had long been renowned for its excellent physicians. Another theory which, with my interest in old breeds of farm animals, had me riveted, was that the call (written down and translated by Ap Ithel), with which the Lady is supposed to have brought all her animals back home to the lake with her, was probably a list of primitive breeds of cattle in Wales:

'Brindled cow, white speckled
Spotted cow, bold freckled
The four field sward mottled
And the grey Geingen
With the white bull
From the Court of the King
And the little black calf
Tho' suspended on the hook
Come thou also, quite well home!
The four grey oxen
That are on the field;
Come you also
Quite well home!'

'The Court of the King,' said Dafydd Ifans as we read this together, 'was Llandeilo, not far from Myddfai. And, as you know, the castle there was famous for its white cattle. And it's very probable, you see, that the King lodged his Court Physician at Myddfai. I'm talking about the time of Rhys Grieg who was the Prince of South Wales and died in about 1233. But later, when the Anglo-Normans were in charge in all of Wales, about the end of the thirteenth century, it says in the Dues of the Lordship of Llandovery, that the ten free tenants of Myddfai must supply a physician to follow the lord at his own expense! So you see there's this strong tradition of Myddfai and doctors going a long way back.'

It was all marvellous stuff which gave substance to the feeling, I always have, about the whole district having something very medieval about it. Even when I'm belting along the A40 between Llandovery and Llandeilo, I sometimes think that it wouldn't surprise me at all to look out of the car and see a troupe of horsemen in gaily coloured robes and jaunty hats, galloping along some distant, winding track between the hillocky and hummocky landscape. Further up in the hills, the feeling is even more intense, if more sombre, particularly in very early spring when the earth is turning over and the fields and trees, still blasted from winter, are just beginning to show the first promise of life. Then, with everything stripped of its finery, but promising more to come, the sense of history seems to shake itself out and the tiny roads, which lead on and round and disappear tantalizingly into the distance like the fairy roads in a child's book, seem only to wait the creak of ancient carts or the clop of hooves. Sometimes, as I stand looking along such a road, I get a shock when nothing more than a farm bike comes skittering along or a great lorry lurches round the hedge.

This sensation of time overlapping reminds me of Daphne du Maurier's book, *The House on the Strand*, in which the hero takes a drug which allows him back into the medieval world as an unseen spectator, drifting centuries from one step to another. In spite of his uncertain end, because of the effects of the drug and the dangers he faced from the modern world (still operating its trains and cars in spite of his illusion of empty space), I envy him greatly, and Daphne du Maurier for being able to write about that world so vividly, so that you can almost smell it. But this double vision of mine is not something many people sympathize with and Myrddin certainly had little time for my ravings about the far distant past. He was much more concerned with what had gone on in the nineteenth century, but he did volunteer one priceless piece of information.

'They say that one of the farms in Myddfai, Llwyn Mereddyd, used to belong to the Physicians,' he said thoughtfully. 'I know the people who live there. Why don't you give them a ring?'

And so I was sitting in the peaceful room with the ticking clock and the strange, low whining of the wind, listening absorbed to Mr Ewart Jones whose family had lived there for generations. As he talked quietly and answered my questions with infinite patience, the historical veracity of the Physicians faded from my mind and it was full once more of the original legend of the Lady, coming back from the lake to instruct her sons in the place they call Pant Y Meddygon.

'Did people really believe in the Lake Lady once?' I asked him.

'Oh I don't know, but they used to go up to the lake every first Sunday in August. You'll see them now!' he said firmly.

'No,' I said sadly, 'just the hikers. Why did the local people go before?'

'Oh, it was a day out. I don't think they hoped to see the Lady. But they used to go and have a picnic there by the lake.'

'Do you believe in the Lady?' I asked him.

He smiled, a lovely smile that twinkled in his eyes, and said:

'Oh no. It was just a story. People wanted to know the beginning of things you know. But it is one of the *best* fairy stories.'

It *is* one of the best fairy stories but, according to several folklorists, there was probably more to it than that. The lake could have been a primitive place of worship to a Lake Goddess, with the ancient pilgrimage to worship there surviving as a folk memory which lasted down to the days of Mr Jones' youth when people went to visit the lake in August. Another, more prosaic scholar, thought that the visit was really to observe the eels swimming on the surface of the lake at that time of year, for the lake and the little river, the Sawdde, which rises from it, were once famous for their eels.

Another part of the legend (which may bear out the pilgrimage theory) concerns those 'four grey oxen that are on the field'. They followed the Lady when she called them, all the way from the farm in Myddfai where she and her husband lived, plough and all, right into the lake, leaving a furrow which, they say, can still be seen.

So, on yet another bitterly cold day, with the wind, not seeking harmlessly round the cracks of a house but left to fly triumphantly free across the mountain, I went with a local farmer, Willy Davies, on a long trek along winding sheep tracks, around massive boulders and squelching bogs and through a dip in the mountain to see those famous plough marks.

'That's it,' shouted Willy above the wind. 'We were told when we were children that these are the plough marks the oxen made. You see how it swerves about, just like a plough going fast and no one to guide it.'

I peered at the ground and there, sure enough, were what, with some imagination, could be mistaken for very drunken furrow marks.

'I thought you could see it all the way to the lake,' I yelled back to Willy.

'Well, they do say you can see it all the way. When the sun's in the right direction or something.'

'And is it true that you can see the hoof prints of the cattle in the rocks by the lake?'

Willy paused and scratched his head. 'I thought it was the ponies' hoof prints,' he cried.

I turned my back into the wind to try and give the microphone a bit more shelter and put the big one to Willy: 'Do you believe in the legend?'

Willy looked around as if the hills could overhear.

'Well I wouldn't like to say I did now,' he said rather sheepishly, 'but it's such an old story, there must be something in it!'

It was blowing a gale again on the day I had to meet Jack Powell up by the lake itself. Only this time it was also pouring with rain, so that water ran off the roads. The river roared deafeningly and the boulders it was pushing in its path sounded like pistol shots as they smashed into each other. The rain gauges registered that we'd had a good inch of rain in the night.

I gleaned that little bit of information from Les Marshall who had looked after the lake in its modern capacity as a water supply for many years. He was rather distracted about the possible damage the torrential rain had done to all the fixtures and fittings which got the water filtered and far away down to the suburban householders, boringly unaware of all the drama taking place up in the hills to get their kettles filled and their loos flushed.

Les had been up for hours, making sure that all was well but, when I arrived at his house on my way up to the lake, had time to give me reams of information about the more recent history of the lake. He told me about the Irish workmen who'd originally been imported to bank up the sides of the lake and make the sluices and lay the pipes which sent the water on its way down to the filter beds, and of the pacifists who'd been sent to work up by the lake during the First World War and left to support themselves on what they could grow.

It must have been hard on them, those intellectuals who'd never done manual labour before, dumped in such a remote place to survive, and I've often wondered if the concert pianist amongst them, his hands roughened by the rocks and the gravel and the cold, ever performed in public again. As I listened to Les Marshall, the Legend of the Lady herself seemed to lose some of its drama, but he had a story about her too. It wasn't in the least bit romantic.

'They made a television film up here about the legend you know,' he said, 'and they wanted a shot of the actress, who was playing the Lady, walking out of the lake. Trouble is, now it's a water supply, no one's allowed to go into it. Anyway, they got permission from the Water Board to do this shot, as long as the actress had been completely washed, in a special disinfectant. And that was my job – to see she had.' He eyed me carefully.

'And did you? Wash her I mean?'

'Well . . . it was a bit embarrassing,' said Les, grinning. 'I took her up to the work shed and handed the disinfectant to her round the door. I put her on her honour to wash herself all over in it, and I stood outside and listened. Anyway, that was the least of her troubles! They had the poor girl hopping in and out of that lake time and time again!'

When Jack Powell arrived a few moments later, Les gave us both a lift up to the lake in his Landrover. We bumped and clattered over the rough track, which winds up from the filter beds, and everywhere tiny streams of water gushed along beside us or spouted out of cracks in the rocks. When we got to the lake it was almost impossible to stand against the wind spiralling and whipping through the great glacial hollow below the Fans. Waves raged and tumbled about on the surface and poured over the sides of the lake and great jets of spray drenched us completely.

'You'd never think,' cried Les above the tumult of rushing water and screaming wind, 'that there are times up here when it's so still the Fans are reflected deep in the water and you get a mirror image of everything!'

Jack Powell snorted and led the way firmly into the work shed and out of the wind. He sat down and calmly surveyed the raging fury outside as if he had all the time in the world.

'Damn thing,' he said, pointing at my microphone.

144

Jack had worked up by the lake for many years, repairing stone walls, clearing out the sluices, mending the roads and doing a thousand other maintenance jobs. Before that, however, he'd worked on the land and for a long while now I'd been hauling him, protesting, in front of my microphone to talk about all sorts of things. Jack could still remember the days when he went to the hiring fairs to get his next job and could describe those jobs and the people he'd worked for with a wry humour I never tired of. It was Jack who'd told me how they used to milk the ewes for two weeks after the lambs were weaned; of the way they used an old ladder placed endways, to halter the ewes by their necks, and then went up and down the row stripping their udders. The milk was used to make cheese for the farm itself, the cow's milk reserved for making butter for the rich man's table. Sometimes traders would come up from the valleys and buy the hard sheeps' cheese for the miners.

I first broadcast this business of milking sheep when I was still working in London, and I remember everyone in the studio pulling disgusted faces. Later, when I heard of a young couple in Devon milking sheep commercially and rushed to see them, the general reaction in the *Small Country Living* studio wasn't much better. It's odd, that in so short a time, milking sheep has become quite the 'in' thing to do, but I always think of Jack Powell when I hear or read of the latest trendy report about the sheep milking business in Britain and hear him saying, 'Damn me, it was a bugger, going up and down under those old ewes with a saucepan getting a few squirts at a time!'

Now, however, it was Jack's years of experience up by the lake I wanted to hear about. I knew I wouldn't get any romantic visions of beauty or the mystery of working in that lovely, legend-haunted place from Jack but, remembering my own wondrous find of a little flint arrowhead from the lake, I asked him if he'd ever found anything there himself.

'Dead sheep sometimes,' said Jack. 'Silly old fools fall off the rocks up there.' He paused and thought. 'We did find a clay pipe once from when the Irish were here, and there was an eel all mangled up in the sluices I remember.'

'What about the legend?' I asked him warily. Jack looked at me, smirked and said nothing. I decided to put it more directly.

'What would you have done, if you'd been working up here alone and suddenly saw the Lady rising up out of the lake?'

Jack looked at me as if I'd gone totally mad and then he chortled: 'Who me? Why I'd 'ave buggered off 'ome!'

Chapter 15

It seemed strangely lonely on the meadow. The river ran its quiet course, a dipper bobbed on a rock and then flew straight along the water, a woodpecker called over in the wood, and far away from somewhere on the hills, came a faint sound of whistling and the protesting cries of driven sheep.

I kicked at the small pile of blackened ashes on the river bank and wondered where Mike and his mule would be by now; somewhere out along the old drovers' road, away over the summer hills, the bells of the mule tinkling merrily against the sound of lark song. Only that morning they'd been here on the meadow, the smoke rising from the little fire in its circle of stones, the mule grazing peacefully and Mike and I sitting on the rocks drinking scalding tea out of old mugs.

They'd come a long way to see me, Mike and his mule Muffin, taking days to wander along forgotten tracks to arrive at the old pink door of the cottage; Mike with his beard and his ancient bush hat,

Muffin bearing all their possessions and bedecked with a cluster of Indian bells which chimed every time he moved. They camped down by the river, refusing all offers of cottage or stable, and the dogs and I took to visiting them there every morning to partake of tea and chat, as the sun warmed our bones and the smell of meadowsweet rose on the air.

It was a way of life Mike had chosen deliberately, this wandering around with his mule during the summer, sometimes hop-picking or doing other seasonal work. In winter they retired to a field near Hay-on-Wye, where Mike had built himself a small primitive shelter. I can't remember how I'd heard about him, or how I'd contacted him in the first place, but coming to see me for the programme to talk about his way of life, had given him the perfect excuse to explore the countryside around the Black Mountain and his arrival provided my neighbours with enough amazement to keep them going for weeks.

'Damn! I thought it was Jesus Christ himself when he came along the road with that donkey!' said one old farmer.

Indeed, of all the people over the years who have come down the steep hill to my gate, none have been as spectacularly unique as Mike and Muffin, an effect which I rather suspected Mike enjoyed a bit too much for one who claimed such indifference to worldly things. But the curious thing was, the burnt patch of ashes where he camped took years and years to disappear, and I still call that part of the meadow, the Place of the Mule.

There are many other places on the farm belonging to certain people. Fred Archer and I always sit on a stone wall, looking out across the valley, to do our annual batch of interviews and that is Fred's Place; just under the bridge is the Place of the Otter That Wasn't, in memory of the keen young man who came to look for otters and brought me a fresh otter spraint so I'd know what to look out for in case the otters ever came back to my stretch of the river; there is Gwyneth's Place where she and I picked holly, and Andrew's Place where he winched an enormous tree trunk out of the river after a storm. But Madge Hooper has many places dedicated to her. There's the Place of Madge's Amazing Discoveries, where she found several wild herbs even the plant boffin hadn't noticed and also some poisonous ragwort for me to get rid of before it spread over the field; there's Madge's Mourning Widow Place, where she finally identified a geranium plant that wandered everywhere and had been puzzling me for years; there's the Place of Madge's Goat Surprise, when five goats emerged from a small goose shelter and we wondered how they'd all fitted in and thought of claiming a record. Above all, Madge's presence is everywhere in the cottage in the form of the loveliest, most evocative pot-pourri I've ever come across.

To see the real Madge however, you must go to the little Hereford-shire village of Stoke Lacey and find her tiny, hidden herb garden.

It was early summer when I first bumped the Beetle down the rough track which leads to that garden. It was hot and I was tired after an early start and a long route dotted by interviews with various people and their animals, and there was still a long way to go before I could stagger into a hotel, check over the tapes and call it a day. The narrow gateway, which I had to yank the car through, and the curved gravel drive in front of the modern bungalow, didn't promise much delight in store. Neither did the slim brisk woman, with white hair and a decidedly no-nonsense expression, match up to my idea of a charming *olde worlde* herb gatherer. She hadn't, I remembered, exactly been over the moon when I'd asked to come and see her. I sighed, switched off the engine and plastered what I hoped was a winning smile on my face. Madge Hooper glared back at me and frowned impatiently. The smile wilted and I tried to replace it with a look of anxious concern which almost got there until a dreadful yawn yanked it out of place.

'I'm so sorry,' I murmured through the car window. 'Long day. I do hope this isn't too inconvenient for you.'

Madge put her head on one side, peered at the ancient car and then smiled stunningly.

'It's Jeanine isn't it?' she said. 'I thought it was some customer ignoring the "Closed" sign on the gate. I didn't want us to be disturbed just now, so I put my cross face on, in case! Come in and have a cup of tea.'

I opened the door of the car, eased my aching legs out onto the gravel path and, as I did so, perfume, subtle and delicate, drifted in a breeze around me and a bird began singing its heart out in a lilac bush. Tension and the mounting cynicism of the day left me and I felt suddenly as if I was coming home, as if I'd known this place and this person all my life and loved them both.

All around me, the little garden gave out its scent and its birdsong and soothed my eyes with its cool composition of trees and shrubs and rioting roses, planted as backgrounds to the bright carpets of jewelled herbs spilling out of their beds, or standing tall in clusters against the walls. Madge moved her hand over a cascading fountain of leaves and crushed a few to release the haunting perfume of sweet briar.

'Smell that,' she commanded, 'and you'll soon feel better,' and she led the way into the bungalow.

It was as if we'd never left the garden. It glowed at us through the French windows and its perfume was there in the lovely old bowls of pot-pourri; not dry, dessicated pot-pourri, but still bright with colour and not needing one whiff of any artificial essence to help it. Madge would scorn to use anything but the lavender, rosemary, old fashioned cabbage roses, eau de cologne mint, mellilot (which smells of new-mown hay) and all the other scented herbs she grows. I have a bag of that pot-pourri beside my bed now and if I wake from a nightmare, I clutch at it and sniff deeply. There is nothing like the true scent of an English garden to dispel the cloying horror and swing the world round to normal.

Madge goes back a long way in the growing of herbs. She studied and worked on a large herb farm before the war, and later set up her own place which was geared to supplying large quantities of herbs to the pharmaceutical industry, before so many of those herbs were brought in, looking tired and dusty, from abroad. All Madge's dried herbs are as green and tasty as the day she picked them. Later, she sold her land and, retaining only a small plot, she built the bungalow and created a scaled-down herb farm to supply plants for other gardeners. In a small shed at the bottom of the garden she sells her pot-pourri, herbal teas, herb ointments and the best pomanders I know; the kind that go on smelling of oranges and cloves for years and years.

When she's not tending the garden, drying and preparing her herbs and coping with a busy mail-order business, Madge gives courses in herb-growing. Just lately, she's found time to write a book, giving all her secrets away; secrets which I began to glean from her on that hot day when I first went to Stoke Lacey and persuaded her to become one of the regulars on *A Small Country Living*.

It was this contact with Madge which finished the work Ann Young had begun on Mrs P, and forced me to find more and more room for the herbs which she instructed me to bring back from my Stoke Lacey's visits, or Madge to bring when she came to visit us. Not that all this turned me into a gardener. Mrs P's green thumb missed me completely. I have a very basic, working relationship with the plants in my garden. I dig them in, give them water and a few encouraging words and let them get on with it. Lots of them manage to thrive, much to my profound amazement as they pop up year after year. But it's not much of a garden. As the radio and television gardener, Geoffrey Smith said, after taking a good, long look at it: 'Actually, with a view like this, you don't really *need* a garden do you!' It's not easy either, with every hen and duck, goat and sheep on the place spending most of their waking hours figuring out how to get into that garden and dig it up or prune it back for me. Most of Madge's herbs got planted in big pots, out of harm's way, and if they were silly enough to escape into the garden proper, that was their lookout.

On my initial trip to Madge's garden, I was not alone. I had Barbara Leach with me. Barbara is by way of being the Welsh counterpart of Muriel Sassoon, able at a glance to pick out the one good piece of china or furniture in a room full of dross, generous to a fault, a passionate animal lover and eccentric enough in her own way to put her high on my list of favourite people. She has, however, one incurable fault. She is one of that breed of fanatic gardeners who simply cannot leave other people's gardens alone.

It was an embarrassing trip, with Barbara sneaking off to see what she could beg, borrow or buy from the gardens of my victims. Not that they minded; they were much happier talking plants to Barbara and robbing their own gardens on her behalf, than doing their interviews with me. None of which did my ego, my time-schedule or the space in my car much good. I began to pray, as we drove along, that the next person would be a lousy gardener like me; the car already beginning to look like a greenhouse with fronds and leaves and roots poking out everywhere. I'd been very worried as we approached the herb garden.

Actually, the first thing Barbara had ever done, when I met her, was to offer me a plant. It had been a bleak, wet day during that supposed working holiday when the Artist and I had originally come to Wales to stay with the Summers. Because of the weather, we'd gone to investigate one of the local towns, which we found empty and deserted except for the forlorn faces of shopkeepers peering out at the empty, rainswept streets. We paused in front of a window crammed with all kinds of bits and pieces; Staffordshire figures, brass candlesticks, old prints, oil lamps, tattered books, ancient spotted mirrors and a small stuffed crocodile all jostled for space, with none of that pretentious effort to make the least look the most which you see in many antique shops. It was just the sort of place to rummage about in on a cold rainy day and the Artist and I didn't hesitate before pushing open the door and jangling the bell discordantly. Somewhere, a wild barking of dogs answered it, but of the proprietor there was no sign.

'Not very nice this weather is it?' said a voice from deep within the glorious muddle and a small person with vivid blue eyes popped out from behind a cupboard, her hands full of roots.

'I was just sorting some plants out,' she explained. 'Would you like a couple? They call this Joseph's coat because it's all colours at once.'

I looked at the tangle of roots doubtfully.

'Well, actually, we're on holiday, sort of,' I said.

'Never mind, have a cup of coffee instead,' said this surprising shopowner.

As that was exactly what we'd been looking for disconsolately in the town for the past half hour, we nodded happily and followed our new friend out through the door and into a dark passage, where we were bounced upon enthusiastically by a delirious gang of Cavalier spaniels and a rather dignified dog, which looked not quite like a Corgi and was addressed by his owner as Jokins.

Jokins and the rest of the pack preceded us into a narrow room with an immensely high ceiling. A coal fire gleamed at us, but any hopes we had of occupying the deep, shabby armchairs in front of it, were confounded by the dogs which leapt up and settled down on them. It was just like the Summers'. But these live dogs were not the only ones in the room. All around the walls and standing on every available space, there were prints and porcelain and bronze figures of dogs. On the old dresser lining one wall, a haphazard mountain of papers and cards threatened to lose their moorings and cascade onto the floor at any moment; a round table in the window, which looked out onto a garden, held a large pot plant, an ancient tin, adding-machine and, shortly, two steaming mugs of coffee. It was not a tidy room but it was, like all the best rooms, cosy and warm and lived-in. The Artist and I were still there when Barbara's husband Roy came home from work.

It was Roy who patched up the odd bits and pieces of chipped or broken Staffordshire, which was all we could afford to buy from them in those first years, like the pair of lions which look as if they've escaped from Trafalgar Square and got cracked in the process, or the tall greyhound with a hare in its mouth and missing one leg. Since then lots of other, perfect things have found their way from the little shop to my cottage, all of them so cheap it's ridiculous and, whether I will or no, lots of plants have found their way to my garden out of Barbara's, for what Barbara has, she gives.

Both of the Leach's are keen gardeners so when, many years later, I suggested that Barbara might like to come with me on one of my interviewing trips because I was going to visit a herb farm on the way, Roy took time off work to mind the dogs and the shop so that his wife could get away. He didn't warn me about her plant-collecting ways.

Until we reached Madge's herb garden, I'd been able to keep a fairly good eye on Barbara and her mounting collection, but with that wonder of scent and colour blinding me, I temporarily became one of the converted. It was Madge's fault of course. She got me properly hooked on the folklore and legend bit. She could quote endlessly, from poets and prophets and peasants, about those herbs, and she bewitched me with their glamour till I had wild visions of getting down to this gardening business seriously and creating my own piece of magic in the hills. I began ordering plants from her like a true fanatic. Meanwhile, Barbara was quietly gathering up pot after pot from Madge's 'For Sale' display.

We must have been a queer sight, the old green Beetle laden down with luggage and plants and two arguing women, as we went on our way to see a couple with Gloucester Cattle (no chance for Barbara there), a breeding centre for rare poultry (which also sold plants and Barbara got locked in the car), and thence across the long, straight

Roman road and on to the other side of Marlborough and my friends Bernard and Eileen Venables.

Bernard, fisherman, author and artist, was to talk to me about rural ghosts. He and Eileen had lived in several old cottages and in the last one they'd suffered from the thieving ways of the witch who'd occupied it in the past. She had a boring habit of nicking Bernard's manuscripts just before a deadline and then replacing them in exactly the same spot months later. In another cottage, whose original tenants had been good, solid, labouring folk, one of them still came back in spirit for no good reason except to stomp across the bedroom, sit down heavily on the bed and drop his boots, one by one. Every night while the Venables worked downstairs, they waited for the sound of the second boot before they could really get on with what they were doing.

Now, they rented a lovely old house surrounded by an estate, with no very obvious ghosts, but where the pheasants were sacrosanct and raided their garden with impunity. It was in that garden that I had second thoughts about all this herb mania, for Eileen had gone down the same road, not realizing what rampant habits things like Sweet Cicely and the various mints have.

Nevertheless, it was a gardener's garden and I was anxious to get Barbara out of it as quickly as possible. The next day, I did the interviews I had lined up (with an old-style gamekeeper and a sculptor who made life-sized wooden pigs which all reminded me of Blossom) and headed the Beetle back towards Wales.

'There's just one more place I'd like to go, though,' I said to Barbara, as we rattled along. 'I'd like to look at some Icelandic sheep.'

Barbara paused in her surreptitious rearranging of my belongings so that her plants could have more space.

'Not going to Iceland though are you? We've got to get these plants home soon!'

I glared at her. 'There's a lady not far from Chippenham (which we are rapidly approaching) who has just imported three of them. Could you stop messing about with those plants for a moment and get your eyes glued to these instructions because they look pretty complicated to me?'

Barbara sighed and took the piece of paper from me. She'd already had several rockets for getting us lost, her attention constantly swivelling to the exuberant cottage gardens lining the lush summer lanes of Wiltshire.

Apart from a few U-turns where no U-turn should be possible, and a couple of merry jaunts in and out of Chippenham when we should have by-passed it altogether, she did manage to get us to the right village and the tall wooden gate flanked by a notice which read 'Ring Bell and WAIT.'

We rang the bell and waited. 'Why,' asked Barbara, 'are we going to look at these Iced sheep anyway?'

'Ice*landic* sheep,' I growled at her. 'They have wonderful fleeces and I've never seen any before and I'm curious to know how and why this lady imported them. And you leave her garden alone, if she's got one!' Barbara pulled a very rude face.

The door in the wall swung open and a tall, breathless woman motioned us inside.

'Sorry to keep you waiting,' she said, 'but I have another visitor who wants to look at the sheep.' She lowered her voice. 'She's interested in *bones*!'

'Why?' asked Barbara, her eyes shocked.

'I'm not too sure really, but it seems she does research on the bones of primitive sheep and they've found some on an island in Scotland. They're quite big bones, so this lady thinks those sheep must have been about the same size as Icelandic sheep, which are a very very old breed, and she wants to have a look at mine. By the way, I'm not going to talk into that thing,' and she pointed vindictively at my tape recorder.

'Mrs Joly you *must*!' I pleaded 'Everyone will be fascinated by these sheep . . .'

· 'I don't care,' she interrupted. 'I'm no good at that sort of thing. Maybe Miss Noddle will talk into it for you.' She turned, as her companion joined her, introduced us and led us out of the wooden gate and back down the road.

'Where are we going?' whispered Barbara as she trotted after me. I shrugged my shoulders, hitched the tape recorder higher, and set off after our hostess.

We all arrived panting, as Mrs Joly gestured towards the three woolly shapes, grazing peaceably on the field.

'There they are,' she cried. ' And they've already been shorn once this year! And look at their fleeces again. And have you seen such lovely colours!'

They *were* lovely colours; a golden cream and a rich chocolate on the ewes and a deep, silvery grey on the ram.

Mrs Joly pointed to him proudly. 'I call him Dougal, from *The Magic Roundabout*. You can see why, can't you! That's Florence there. Look at their little short tails!'

I stood there enraptured. I understood perfectly, without asking, why Diana Joly had gone to all the trouble and expense of importing these sheep. With their delicate faces and amazing cascade of glowing wool, at that moment they were the one thing on earth I wanted to possess.

I came out of my trance long enough to chase Diana Joly all round the field and try to trick her into doing an interview; to stand at last to talk to Barbara Noddle and ask about the research she was doing; to stop Barbara Leach gazing speculatively at a little patch of marsh marigolds at the edge of the field, and finally to get some real Icelandic 'baas' on tape. At last we all trooped back to the house for tea.

'If you ever have any lambs for sale,' I said to Mrs Joly as we marched, 'I'd be very interested.'

'Heavens! I won't have any pure-bred ones for ages,' she cried, 'but there's something in the garden you might like to see.'

As we entered the big wooden door again, I looked around the neat layout of enclosed yard and buildings and the remains of a garden being thoroughly decimated by a large black ewe and two small black lambs.

Diana Joly waved at them. 'Dougal is their father,' she said. 'The mother is a Black Welsh Mountain ewe. Didn't you say you have Black Welsh Mountain sheep? This one has got mastitis, so I'm having to bottle-feed the lambs, but she minds them the rest of the time. I call them Francis and Panna.'

And thus it was, that the Sagas came into my life.

Chapter 16

They arrived when they were five months old and their tight little astrakhan curls had begun to unfurl into long lustrous locks. They sat, contentedly cudding, in the back of Diana Joly's old estate car, while an eager sheepdog bounced expectantly on the front seat.

'I brought him to practise on your sheep,' explained Mrs Joly. 'All of mine know him too well you see, so he needs strange flocks to test him. What a lovely place!' and she stood on the yard surveying the valley laid out under the clear autumn sky.

'He won't get much change out of mine, I'm afraid,' I told her. 'They treat most dogs with contempt! Still, it will do them good to know how the other half lives. They're so spoilt, I sometimes think they've forgotten they're sheep at all. I do hope these two know their place, I mean you haven't been too kind to the ram lamb have you?'

It was something I had warned her about, for whereas bottle-fed kids grow up into much nicer goats than the ones reared on their mothers, lambs usually grow up into utterly delinquent sheep. They bully and push and demand and are totally impervious to any kind of snub. It's bad enough with a ewe, but a pet ram can be very dangerous.

Mrs Joly waved aside my doubts with an impatient hand.

'These two are very good,' she said firmly and demonstrated their amiable dispositions by leading them across the yard and into the pig-sties which I'd prepared for them with deep beds of straw.

Apart from their colour, they bore little resemblance to their solid, Black Welsh Mountain mother. Their faces were fine-boned, soft to the touch and looked highly intelligent. Their tails, short, neat and tapering to a point on the end, gave their behinds a slightly hitched-up look, rather like llamas'. The long, soft wool and delicate little feet marked out their Icelandic connection completely and I was bursting with pride to think that they were mine.

I'd had a bit of trouble persuading Diana Joly to sell them to me but I was sure somehow that my number was on those lambs and had finally worn her down. Apart from their beauty, I had a very good reason for wanting Francis and Panna.

For some while, I'd been less than happy with the fleeces on my black flock. It is rare, actually, to find a true black sheep and my own had a tendency to develop a rusty appearance with age, or, even worse, go a kind of pepper and salt colour, a sure indication that their wool was full of coarse kemp fibres. I do have one ewe who has kept her coal-black fleece through four shearings, but it's unusual enough for her to have earned the name of Mrs Black. I was not so much worried about the colour of the fleeces, however, but about the texture which, in my flock, felt more like a decrepit pan scourer.

I'd flirted with the idea of putting rams from other breeds with them to improve this fleece texture; Ryelands, those chubby sheep from the Herefordshire borders who once had the privilege of providing fine wool for Queen Elizabeth the First's stockings; the extraordinary Wensleydales with their great cape-like fleeces sweeping the ground and having, as a breed, the advantage of eliminating kemp from the wool of their offspring when mated to another breed. Wherever I went on my travels, I looked at other possible rams and certainly whatever breed I chose, I should have had a pure-bred ram, but those Icelandic sheep thrilled me as no other breed had done. The fact that there were two lambs already with a Black Welsh Mountain dam seemed a strange and most significant coincidence. As always, I'd galloped off down the path signposted by that coincidence and became obsessed with the desire to own Francis and Panna.

If my crossing programme has been less than successful from the fleece point of view (I really *should* have waited for a pure-bred Icelandic ram), I don't regret what I did. Diana Joly's triumphant arrival at the farm was a tonic for a start, her enthusiasm pouring out of her as she darted round the fields or charged ahead of me along the river bank. In the cottage she peered through the little windows and raved about the view from each one and won Mrs P's heart completely by appreciating her lunch with endless compliments. If my sheep were hardly amused by being rounded up and penned by her sheepdog, it did them nothing but good to do as they were told for once.

Diana Joly left, bearing a cheque, one of my own Black Welsh Mountain ewe lambs and Annie, Little-Nana's daughter. As soon as she'd gone, I hastened back to fuss over Francis and Panna.

I've been fussing over Panna ever since, for she gave the lie to that business of bottle-fed lambs growing into monsters. She has been the soul of charm and politeness all her life and is so easy to handle, that I can walk up to her when she is lying down and trim her feet without her even pausing in her peaceful cudding. She's a perfect mother and produces her lambs invariably at the most convenient time (about four o'clock in the afternoon), and gives birth so easily that I've never had anything to do except make sure she is comfortable. It's not her fault that I hover over her hopelessly for weeks before she lambs; it's because of the enormous, milky udder she develops weeks in advance and which always convinces me she's 'going bad' (as they say around here when a ewe is about to lamb), long before her due date. It's something all the Icelandic crosses have inherited, this big, perfectly rounded and very productive udder. In fact, if I ever felt that way inclined, I could milk them all as easily as I milk the goats, and several of them have four working teats which is useful when they produce triplets, as they so often do.

If my initial reaction to Panna proved to be well-founded, my delight in her brother Francis was a mistake. I might have been warned by the bold, speculative look in his eye, that here was the kind of forceful character I could do without. While he was small he was fairly harmless, but as he grew, he proved his horns on everything within reach, especially my shins. My first mistake with him was not to remove Panna immediately. He got her in lamb in just the few days they were together, although I didn't know that till the pregnancy was well on its way and too late to do anything about. Apart from the fact that they were so closely related, at five months old, she was far too young to be in lamb. They'd been late lambs themselves and she needed all the next year to catch up on growing.

I re-named her brother Franco, to match his increasingly irritable character and allowed him, when he was six months old, to serve just a few Black Welsh Mountain ewes and the Potters.

'I can understand why you call the Icelandic crosses, the Sagas,' said Carry Naylor, as we stood watching the Herdwick sheep in their pen, 'but I don't see why you call these the Potters.'

'Beatrix Potter!' I said. 'You know she finally settled in Cumbria and went into farming? Well, the sheep she bred were Herdwicks and she eventually became a highly respected authority on the breed.'

Carry looked at me doubtfully. It had been a dramatic day, one way and another and far from the peaceful, re-charging of batteries which she and Andrew had hoped for. Apart from anything, there'd been the mother and father of a gale the night before, which had made them finally understand my neurosis about the wind. In the morning, Andrew had been coerced into climbing up on the roof of the poultry shed to try and nail down the flapping sheets of felt which had come adrift. After lunch, I'd proposed a little drive to see some people who owned a flock of Herdwick sheep. They lived not far from the tiny village of Bethlehem, which dreams away in obscurity for most of the year, but has a hectic few weeks around Christmas time with people rushing to post their cards at the village post office to get them stamped 'Bethlehem'.

As we bowled along the quiet road which leads to Bethlehem, the pretty, hillocky countryside on either side of it engrossed the Naylors but, up ahead, I saw a horse and rider trotting rather erratically towards us. I kept a wary eye on them, my foot hovering over the brake pedal. Noticing a small lay-by, I began to ease the car into it, to give them plenty of room but, as they drew alongside, the horse swung its hindquarters round and aimed a couple of deafening kicks, broadsides, at the green Beetle. From inside, it sounded as if the whole car was being split down the middle.

Andrew and Carry, jolted out of their rural dreaming, went white with shock and I just sat there, unwilling to get out and look at the horrendous dents which must have been on the wings and bonnet of my poor car. The rider of the horse, meanwhile, was desperately trying to control her mount which was still hopping around like a mad thing. I could see, as it pranced by, that its hooves were garnished with bright green paint. I couldn't bear to think what the car looked like.

At last, the horse tolerably under control, I crept out of the door. There was nothing, absolutely not one single mark on the car.

The horse's owner and I carefully examined its hooves. There was the green paint, but the Beetle presented us with a gleaming, perfect surface. In spite of Andrew, recovered enough to be furiously indignant, even climbing right under the car, there was not a scratch to be seen. How the paint had got onto the horse is a mystery that was never solved.

The Naylors and I were still exclaiming about the strange accident, when we arrived at the Somerfields and nothing would do but the whole family had to inspect the unblemished car once again. We almost forgot about the sheep altogether.

'I've penned a couple for you over here,' said Nick Somerfield at last. 'Just a few old girls I'll be sending off to the market soon.'

Fatal words, with Carry, appalled, exclaiming beside me, 'You mean for slaughter?'

Nick gave her a deprecating shrug and led us to the yard where a small band of defiant Herdwicks stamped their feet at us.

'What a temper!' cried Carry. 'Shan't Sheep, that's what they are! Shan't! Shan't! Shan't!'

If you've never seen one before, a Herdwick comes as a bit of a shock, with a white head, fat, round little ears, huge eyes, and stubby legs topped off by a mass of blue-grey wool. Herdwicks are the great survivors of the harsh winters on the Cumbrian fells.

'. . . through tempest, frost or heat, we live our patient day's allotted span. Wild and free as when the stonemen told our puzzled early numbers; untamed as when the Norsemen named our grassings

161

in their stride. Our little feet had ridged the slopes before the passing Romans. On through the fleeting centuries, when fresh blood came from Iceland, Spain or Scotland – stubborn, unchanged, *unbeaten* – we have held the stony waste.' So Beatrix Potter, in *The Fairy Caravan*, has the Herdwick sheep describe their life and history. It is said of them that they are so fiercely independent, that if they're sold locally away from their own farms and sheep-walks on the fells, they will find their way back, but when the Somerfields, on a trip up north, fell in love with the breed and brought some back to Wales, they had settled well enough.

'You'll probably get a couple of good crops of lambs from them,' said Nick as I eyed the small group in front of us. 'Lord knows how old they are. They were pretty ancient when we bought them, but we put some to the Welsh ram and some to a Herdwick and they've done us proud. Which is why I have to get rid of a few. We're already a bit over-stocked.'

As he spoke, I found my attention riveted by one very old girl whose big eyes gazed back philosophically at me, resigned to what the fates would throw at her this time. Looking at her, I could almost feel the keen blasts rushing along the Cumbrian fells. It was as if all the wildness and beauty of that lovely country was mirrored in the eyes of one tired old sheep.

'How much do you want for her?' I asked Nick.

'Old Number 10?' he said. 'Funny you should pick her out. I don't know if I'd have really sent her to market. She's a great old character that one! I'll let you have her for eleven pounds . . . that's less than I'd get for her for meat, because I'd like to think she's gone to a good home to end her days. I'll sell you a couple of others at the same price to keep her company.'

'Done!' I cried and Carry said, 'Thank Heaven!'

Nick delivered them the next day and, while the others continued to stamp and swear at me, Number 10 merely sighed and gazed about her and waited for her doom. When that turned out to be a bucket of nuts and a full hay net, her whole body seemed to expand and she actually deigned to eat from my hand. It became obvious, as time went by, that Number 10 had once been a pet sheep and, dimly, the memory of better days came back to her. It must have been a long time since she'd lived near the house and been fed from a bottle and bucket; a long time since she'd been turned back up on the hills with the rest of the flock; a long time of living and lambing on the fells; a long time since, of being herded down from those fells, taken to market and prodded and poked and finally loaded up and brought the long weary, jolting journey to Wales. That it should all end in being virtually a pet sheep again, was a fate no one could have hoped on her.

She was very old, that much was obvious, so, instead of taking her back to be mated with the Herdwick ram, I let young Francis serve her and the others, knowing that the lambs would be fine-boned and easy to produce. I called her Hetty, after a dog which Imogen Summers had once rescued; an old, grizzled terrier which had escaped time and again from the dogs' home and made her way back to London by hopping on the train. Somehow, Hetty the dog had that same resigned, but determined look in her eyes which was there in Number 10's, as if they'd both seen a lot of life and didn't mind; they'd survive it somehow. Hetty the sheep learnt her name very quickly; so quickly that I wondered if she'd had a name similar in sound before. Whenever I called her, she answered with a deep, resonant baa like a foghorn; a sound which brought the rest of the flock to a standstill the first time she produced it. You could hear it from one end of the farm to the other.

The Sagas and the Potters (Panna and Hetty in particular) filled a very big gap in my life and in the general atmosphere of the place, for Doli was no longer there to exasperate and amuse me with her large, emphatic personality.

After her failure to get in foal by Wishful Wanderer, Doli had returned from her stay with Paul and his horses and promptly started to produce an assortment of ailments for me to panic over. None of these had any basis in reality, but were obviously an indication of her loneliness and boredom. For years, my friend Sedley Sweeney at the Smallholder's Training Centre near Brecon, had been trying to prise her away from me and at last, with a tear in my voice, I rang to tell him that he could buy her.

'On condition,' I told him, 'that if you ever decide to sell her, you give me first refusal on buying her back!'

'Of course,' agreed Sedley, 'but I think you're doing a wise thing you know. She needs work and I've got plenty of that for her here.

I'll train her to the plough and she'll be very useful as a demonstration horse for the students.'

'She *can* be a bit unreliable,' I warned him, but Sedley was quite sure he'd be able to calm her down when she played up.

The day she left, I was numb and nothing Mrs P, or anyone else, could say lifted my misery. Only the remembrance of Bertie's fruitless journeys (and their cost) and Doli's own depression as she stomped about, stopped me from ringing Sedley and demanding her back. As it was, I took ages to cash his cheque, because until I did, I still felt somehow that she was mine. For a long time it was more than anyone's life was worth to mention her name.

The one consolation was that, without Doli's massive hooves churning up my few fields, I could indulge myself with a few more sheep and that was how I'd succumbed to the Herdwicks.

It wasn't a bad winter, as winters go, that year and by the time lambing came round again, the sheep were fat and contented and settled happily into the hay shed maternity unit. Only Franco was disgusted when I took him away and put him back in the pig-sty. By now he was a magnificent creature, his black wool literally gleaming and his horns curling dramatically round his handsome face. His character, however, had not improved and he'd taken to bashing the ewes off their feed. He spent the next few weeks demolishing every single bucket I put in with him and was much more of a nuisance than dear old Gandhi had been in the same place. Rapidly running out of buckets and patience and nursing endless bruises, I begged Myrddin to let him live with his rams until I dared to allow Franco back with the ewes after lambing. The first thing Franco did was totally to demoralize Myrddin's very stern and utterly uncompromising sheepdog, Ken, by chasing him out of the field.

'That ram,' said Myrddin in a shocked voice, 'is going to be a lot of trouble if he isn't sorted now. Ken won't go near him! Let's hope the other rams, who are a lot bigger than him, will teach him a lesson he won't forget!'

Meanwhile, his sister Panna, pampered, watched over and snug in her own private lambing quarters, gave birth to two lambs, so tiny that I could hold both of them in my cupped hands. Fortunately, in spite of their chronic in-breeding, there was nothing wrong with them and Flea and Fly, inheriting their mother's calm personality

164

and her long, delicate looks but not her lovely fleece, have since given birth to their own exquisite lambs several times.

Hetty and the other Potters, having been put to the ram rather later, were not in the lambing shed, but out on the home field, where I calculated they still had a while to go before I needed to bring them in. None of them had anything much in the way of an udder as yet.

There was a thick coating of snow on the Fans but the valley was a deep lush green and the woods were full of birdsong on the day Hetty produced Sadie. I thought I was dreaming as I looked down the field at the contented humps of sheep basking in the warm spring sunshine and saw a small black shape wobbling around Hetty. Not so small when I flew down the field to see it; a huge lamb in fact and already cleaned up, fed and ready for anything. Hetty stood there worshipping it and giving out her deep baa over and over again with pride. Heaven knows how many lambs she'd given birth to in her time, but it was obvious that Sadie was to be the adored child of her old age. Sadie was never ever allowed to stray more than a few feet from her mother's side; there was none of this leaving the lamb to sleep while the ewe went off about her grazing. If Hetty moved, so did the lamb.

As Sadie grew older, she often tried to sneak off to take part in all the mad games the other lambs were playing, but Hetty's desperate cries soon brought her back, and on one occasion when the lamb gang had gone off down into the woods and Sadie with them, Hetty came, in a state of total hysteria, to find me and cried mournfully at the gate till I went and fetched her lost one. But later, when Hetty herself had an accident down in the woods, it was Sadie, now a hefty adolescent, who came to find me and never left her mother as I fought, and won, a desperate battle for her life.

Their story had a strange ending and I still ponder over it.

I hadn't intended to let Hetty have any more lambs, but, by the time Sadie was old enough herself to go to the ram, she and Hetty fretted and complained so much at being separated that, in a weak moment, I gave in and let the old girl into the field where the mating bonanza was in full swing. In spite of her age, she was in wonderful condition and I thought I'd let Nature and Hetty decide whether or not she should lamb again.

Nature and Hetty surpassed themselves. Normally Herdwicks

only have one lamb; on their native fells one lamb is more than enough for any ewe to look after. Hetty and the other Potters had still only produced the one lamb apiece for me in spite of the better living.

Next spring, Hetty produced not one, but three lambs; big, beefy lambs too, none of them weedy and wanting as so often happens with triplets. Sadie was just as delighted as her mother with the lambs, although she herself was still a while off her own time.

At first it seemed that Hetty and Nature had known best and the old ewe cared for her lambs and looked in blooming health herself. Even so, I helped her with the feeding by giving the lambs extra bottles of goat's milk and made sure Hetty had more food.

Forty-eight hours later, Hetty couldn't get up. Too late I pumped calcium into her; too late I realized what a dreadful strain those three big lambs had been on the resources of that old body; too late I screamed for Bertie to come and work a miracle; too late I sat up night after night making sure she was moved every hour and feeding her the Leo-soup. Hetty fought too; fought with everything she had. There was none of that listless giving up which sheep (and goats) are famous for as soon as they feel a bit off-colour. Hetty was determined she would live and so was I, if willing could make it so. But Hetty grew weaker and weaker and the deep voice fainter and fainter.

I had her alone in the lower of the two loose-boxes and next door I put Sadie who was now about to lamb herself.

'You know,' I said to Mrs P on one of my rare visits into the house, 'I've got this awfully strange feeling that old Hetty won't give up until Sadie lambs.'

Mrs P looked at me oddly.

'What you need,' she said, 'is a good hot meal.'

'No time,' I said, 'I must go and feed Hetty's lambs or they'll start tormenting her.' I'd left her lambs with Hetty because every time I took them away she used up precious energy fretting after them.

When I got back to the loose-boxes, I peered over the half-door of the top one and there, struggling to its feet beside Sadie, was the most lovely lamb. It was, if my eyes weren't quite deceiving me, a pale golden colour, shaded to deep apricot on face and legs. How black Sadie and the black ram had produced such a colour I shall never know. I took just enough time to rush in and check that it was a little female, before returning it to an anxious Sadie and going in to Hetty.

She was sitting with her old head raised and a faraway look in her eyes, as if she was listening to a distant sound.

I sat down on the bale of straw in front of her and took her head in my hands. 'Sadie,' I told her, 'has just had the most beautiful lamb I've ever seen.'

Hetty sighed and gave one deep, deep baa. And then she died.

I called the new lamb Hetty. It seemed the right thing to do, somehow.

Chapter 17

The dull, persistent booming had been getting on our nerves for hours. It was a still day and the sound carried monotonously from the army range far away on the Epynt Mountain.

'They must be using the big gun,' I told Mrs P.

'I'll give them big gun if they don't stop!' she snarled, irritably, 'and if you think I'm going to cook that rabbit, you can think again!'

'What's the matter with you?' I said. 'You've been in a bloody mood all day!'

'I don't like bangs, and I don't like rabbit,' she said firmly and she picked up Winston, marched out of the kitchen and plonked herself in the armchair by the fireplace next door.

I looked ruefully at the neatly packaged rabbit haunches I'd brought back from town, thinking they'd make a bit of a change, although to be quite honest, I'm not too keen on rabbit myself. If you've been brought up in a country where rabbits swarm in their millions and if you've seen the horrors of those rabbits smitten with myxomatosis, you tend to go off rabbit. Having since seen the way the tame ones are reared intensively in horrible, constricting cages, I no longer buy it anyway and I can live without chicken too, unless I'm a hundred per cent sure it's been reared free range.

I put the kettle on and made a cup of tea in the giant pot which is just about big enough to cater for our Australian tea-thirst. Both of us have half-pint mugs and the tea has to be really strong.

When I handed her her mug, Mrs P was deep in the latest bit of blood-and-thunder she'd found on the library van, and Winston was snoring loudly on her knee and almost drowning out the distant thuds from the range. My mother looked up at me over her reading glasses and smiled apologetically.

'You know what it is. It's all to do with the outside dunny and the Rabbitoes. The banging and that rabbit brought it all back. I've told you the story before.' She looked at me thoughtfully.

'Tell me again,' I said with a grin.

My mother took a long sip of tea, removed her glasses, patted Winston and put on her story-telling face.

'Well, you know, as a family we had a pretty up and down time of it . . . one day good schools and the best clothes and the next glad to get a roof over our heads and praying the landlord didn't come for the rent too soon. It all depended on whether your grandfather was gambling or not. Anyway, there was one time, when I must have been about five or six, when things were really bad. We were living in a few rooms in a house which stood in a row. Everyone there had to scratch for a living. The only lavatory was an outside dunny that stood back-to-back with the one next door. It was a bit difficult for us kids, with so little room, and so I used to go down the garden to that outside dunny and sit on the big wooden seat and pretend that the planks in the wall were children and play school. Used to have a lovely time, being the teacher and singing away to myself.'

She paused and took another long sip of tea before going on.

'Next door to us lived a family of "Rabbitoes". They called them "Rabbitoes" because they used to walk up and down the street yelling "Rabbitoes! Nice fresh rabbitoes". Of course that was in the days before myxomatosis and these men used to go out to the country and get as many rabbits as they could and come back and sell

170

'em for about sixpence each, all skinned. They sold the skins separately you see. Well, I'm not too sure how they caught the rabbits, but they must have blown the warrens up or something because they used explosives. What we didn't know, was that they stored them in their dunny along with a lot of the rabbit skins!' There was another dramatic pause before Mrs P continued.

'Anyway, there I was, sitting on the dunny seat, singing away to myself and the planks in the wall, when there was this terrible bang! And the next thing I knew, I was flying through the air with a lot of muck and the sky raining rabbit skins and bits of wood!'

I knew the story well, but I never failed to laugh at this point. The picture of the infant Mrs P in her frilly bloomers, sailing through the air pursued by a flight of rabbit skins has always been too much.

'But I can't understand how you weren't blown to bits,' I spluttered.

'Neither can I because both the dunnies were, but I did have a sore behind, I remember that. And do you know, your Nan gave me a hiding afterwards for being in the dunny so long. I think she got a fright as a matter of fact! Not half as big a one as I did though. Anyway, I never liked loud bangs and rabbits and outside dunnies after that.'

'Well, you've never had much luck with outside loos and that's a fact,' I said. 'Do you remember the one on that station, the other side of Moree, which was miles from the house and they had a pet kangaroo which used to go and lean on the door and stop you getting out again!'

'The kangaroo was all right, it was the other place with the turkey which used to bail me up that I didn't like. And there was the one which had all those cows that used to follow me and then scratch their backs on the outside of the dunny till it fell over!'

I shook my head at her, remembering all the years of rescuing Mrs P from the small tin sheds, usually placed a good, long way from the farmhouse, and all the livestock who were determined to keep her in their midst. It was the one thing she never liked about our Outback wanderings, the sanitary arrangements. Even when I first came here, there was (and still is) the Welsh version of our Australian dunny, the tŷ bach, and Mrs P had had to be rescued from that twice; once from a group of skittish heifers grazing nearby and again from Douglas and Daisy the geese, as they demanded to know what she was doing in there anyway and kept pecking at her legs through the door which never would shut properly. She wasn't really happy about the farm until I had a proper, indoor loo.

Not that that didn't have a bit of a chequered history too. It relies on a septic tank, which lies hidden out the back, and all plumbed together originally by a builder with a bit of a hate on the Council Planning Department.

'They won't tell me the right way to make a septic tank,' he'd snorted contemptuously, when we asked if the system he used was the same as the one on the plans. 'I've got the best of those buggers before and I'll get it again! Anyway, you can use your lavatory now. Flushes perfect it does!' and he pushed the handle down with a flourish and left us sighing with relief that the long journeys through the mud to the tŷ bach were over.

A few days later, the Council Inspector came to look at the septic tank and informed the builder in person that he wouldn't pass it unless it was done exactly according to the plans. And this time he won.

The Artist and I didn't really mind stomping in our wellies down the garden, across the yard and into the field to use the tŷ bach again, because the builder, who'd bullied us relentlessly and made our lives a misery for weeks with his rough and ready ways, could be heard loudly swearing from the bottom of the septic tank. He was down there dismantling bricks, wreathed in loo paper and worse, for the Artist and I, on his say so, had indulged ourselves over the past few days, and if ever we had our revenge for the years we had to spend patching up his work, it was then.

Since that time, I'd more or less ignored the workings of the septic tank, apart from making sure that nothing too bizarre went down the plumbing, and keeping bleaches and such things to a minimum. If I wondered sometimes about having it emptied, my friends with a similar system told me to forget it.

'If it's a good one, you shouldn't have to bother,' they assured me, and one said, 'Mine hasn't been touched for twenty years, at least!' But after my tank had been there for a fair old while, I began to notice a bit of a nasty smell tainting the summer breeze. One afternoon, I rang the Council.

'Well now,' said the helpful man from the Health Department, 'you should really have them emptied every year or so. You say yours has been untouched for about ten years? It's probably seeping muck out by now. Could be quite nasty if you've got animals grazing that field. I think we should send the lorry up quickly. Would two o'clock tomorrow do you?'

'Yes. But there's just one thing. If, as you say, people should have their tanks emptied regularly, I'd like to do a recording with your chap when he comes. Lots of my listeners have septic tanks and it might be a good idea to remind them that they could be a health hazard.'

'Fine. I'm sure he won't mind. He's got a job to do in the morning, but he'll be with you straight after he's had his lunch. Make sure he tells you about the chickens!'

'The *chickens*?' I cried.

'Yes chickens. He went to clear a septic tank once and he couldn't understand why the pipe kept blocking up. What it was, was a lot of chickens. Dead, of course, you understand! So he asked the farmer what the heck he was doing chucking chickens in his septic tank. And do you know what?'

'I can't bear to think,' I said faintly.

'Well it seems the builder who put the tank in, told this farmer that to get the bacteria and everything going properly, if he had a dead chicken or something, to put that in. So the silly fool used to go and kill a chicken every week and feed it to the septic tank. Must have cost him a fortune and it didn't do the septic tank much good either. It was full to the brim with dead chickens, feathers and all!'

He was still chuckling and chortling to himself as he rang off.

'Nasty,' said Mrs P when I told her the gruesome tale. 'But did you say they were coming tomorrow? What about that chap from the phones coming?'

I stood there horrified. The 'chap from the phones' was no less a person than the head of British Telecom for South West Wales. He was coming to do a few interviews about rural phone services and, in particular, the way British Telecom were thinking of cutting down the number of phone boxes out in the country.

'Much better if I come out there to you,' he'd said, when we discussed the matter. 'Here in Swansea I only see the urban problems. I'll look forward to seeing a few rural telephones for a change.'

Now, it seemed, Mr Barnes might be seeing (and smelling) a lot more than telephones. He had said, however, that he'd have to leave immediately after lunch to get back to Swansea in time for some important function, so I figured he would be well away before the septic tank lorry arrived.

The day dawned bright and clear and gave no promise of the untold dramas ahead. Punctually at 10.30 a smart, well-polished car drove into the yard and from it emerged a tall, beautifully groomed man in a lightweight tweed suit. He smiled at me expansively.

'What a good idea to come,' he said, 'much better than being interviewed in a stuffy office. The countryside round here is magnificent!'

'Coffee?' I suggested, 'then we can have a chat about what we want to discuss on tape.' I led the way into the cottage where Mrs P had fresh scones ready and had put the coffee into the only cups I possess.

It was all very civilized and the only interruption was when a friend of mine, called Gay, arrived on her pony to get on with some whitewashing for me.

Gay and her husband ran a pony-trekking centre further up the valley and, during the season, were frantically busy. That year however, things were a bit slow getting started and, when I'd been moaning about the advent of another radio series and how was I going to get time to do all the summer jobs like whitewashing, Gay, who had incredible energy, offered to give me a hand. The day promising fair, she'd decided to snatch a few hours to come over and make a start.

I left Mrs P to entertain our visitor. While Gay put her pony onto Doli's old field, I got out the bags of limewash, the brushes and buckets and ladders. When she rejoined me, I explained to Gay how important it was that everything was to be kept quiet for the next few hours while Mr Barnes and I did our interviews.

'We'll be coming out onto the fields to do one about watching out for underground cables when you're ploughing, and we'll probably go off to do something about the phone boxes,' I told her. 'So if you see anyone coming, head them off will you?'

'Everyone?' asked Gay. 'What about them?' And she waved her hand towards the gate. I glanced across and there, huge, yellow and uncompromising, was the septic tank lorry. I looked at my watch. It was just coming up to 11 o'clock.

'Hullo then!' cried the small, wiry man who jumped down from his cab. 'Thought we'd do you first after all!'

'I don't want you to do me first,' I wailed. 'I've got someone else here and . . . Oh Lord!' I turned to Gay. 'Just keep him occupied will you, while I go and warn Mr Barnes!'

Mr Barnes couldn't have been nicer about it.

'Plenty of time,' he assured me. 'Don't worry. Your mother and I have been having a fascinating chat about the days when she was a Bush Nurse and had to use the pedal radio to talk to her patients and the doctor.'

'I'll just shut the windows, though,' I told him. 'They say there shouldn't be any smell, but just in case . . .'

I left Mr Barnes and my mother hermetically sealed in, collected the tape recorder and rushed back to the waiting lorry driver.

We had a few awkward moments getting the lorry swung round on the yard, but finally it was backed up near the septic tank, the big pipe unwound and in position and the driver and I waiting near the cover of the tank.

The ensuing event of the cover being raised, choking cries of horror all round, the sound of the great pipes cleaning out the tank and the pithy comments of the driver and his mate about this tank and others, much worse, they'd cleared out in their time, made a wonderful piece of radio. The only thing I forgot to ask them about was the chicken story. As for the smell, there was virtually none at all and soon, with the pipes back in place on the lorry, the tank lid closed down firmly, the council's form signed and the two men back on the lorry, the whole exercise was over. I waved them a cheery goodbye and returned to Mr Barnes.

'Well,' I said, 'we should have a completely clear run now. About this business of phone boxes. Far from taking them away, I think British Telecom should put more in! I mean here, for instance, people are always coming to ask if they can use the phone to ring the AA, and all those kids they dump out here on these adventure schemes, like the Duke of Edinburgh's Award, are forever popping in to ask if they can ring up someone and tell them they're lost! It's a pain in the neck!'

'Oh, come now,' smiled Mr Barnes. 'I'm sure you're exaggerating. How many people do you get in a year? One? Two?'

I was just about to answer indignantly, when there came a delicate tapping on the door. It was Gay, her overall, her long plaits and her glasses generously streaked with whitewash and an apologetic smile on her face.

'Sorry to disturb you,' she said, 'but the men from the lorry wonder if they can use your phone. I told them you wouldn't be pleased, but they're in a bit of trouble.'

I glanced at Mr Barnes triumphantly. 'See what I mean?' I said and I could tell he thought we'd set the whole thing up.

Gay stood aside to let the two men in and they immediately broke into explanations: 'Sorry bach, but the lorry can't get up the hill and we can't raise them at the Council on the radio.'

I nodded. For some reason, the valley is a black spot for two-way car radios.

I was not too bothered at first about the conversation going on between the two men and the Council. Mr Barnes and I smiled at each other politely but I saw him glancing surreptitiously at his watch.

'Well I think the bloody crank shaft's gone,' cried the driver into the phone. 'You'll have to send someone out. She's stuck on the hill and we'll need a tow in. We'll have a bit of lunch while we're waiting!'

'Perhaps we should do the same,' I said to Mr Barnes, as the men left, still apologising. 'Then we can go like lightning and get these interviews done and you away to your function.'

The lunch, which Mrs P had gone to infinite trouble to prepare and present as only she can, was not peaceful. There was an urgent call for Gay from her daughter, the septic tank operatives came trotting back at least three times and finally the engineer himself came in to have a heated discussion with his bosses back at base.

I must say Mr Barnes bore it all with remarkable patience, especially when I kept pointing out that a public phone box on top of the hill would have saved us all this frantic coming and going. Time, however, was flitting past and still we had put not one word on tape.

Finally I suggested we forgot any travels further afield and simply did our interviews on the yard to get a bit of rural background.

Mr Barnes was very lucid, amusing and collected when I interviewed him, which says a lot for his style. Gay's pony insisted on standing near the fence and blowing and stamping furiously at the flies; there were sounds of revving and swearing from the direction of the lane and the sight of Gay, by now almost completely whitewashed herself, trotting backwards and forwards with her brushes and ladders, was a bit distracting. At last she tip-toed past with a saddle, collected her pony and went away up the field. After each stop to allow either for the passing of Gay, or for me to yell over to the men to keep quiet couldn't they, Mr Barnes picked up the thread and went on with his interview beautifully.

'I think that's it then,' I said, as I switched off the machine for the last time. 'I'm terribly sorry about all this. How about some eggs to take back with you, just so you'll feel you've been in the country?'

'Thank you. I really must get away now. I have about sixty people waiting for me at this function in Swansea.' Mr Barnes eyed his watch with a worried frown. 'If I go like mad, I should just make it!'

Hurriedly I put some eggs in a box, collected a few flowers from the garden and tucked Mr Barnes up in his car. He let in the clutch and eased down the yard towards the gate. The men from the Council were just opening it to allow the septic tank lorry to roll back and settle, bang in my entrance.

'What the hell do you think you're doing?' I shouted.

'Well, we were wondering,' said my friend the driver appealingly, 'if we could leave her in your yard tonight. We can't get her up the hill.'

'But there's a lay-by at the bottom of the hill you could have rolled it into,' I cried. 'If you leave that lorry here no one can get in or out!'

'Yes, but someone might steal her if we leave her out on the road,' said the driver, his face creased with worry.

'Steal it! With lord knows how many tons of filthy muck on it. You must be mad!'

The driver's face puckered. 'They've already scratched her,' he said sorrowfully.

'They' were the engineer and a couple of assistants who'd been sent, with nothing more than a Landrover, to haul the huge loaded lorry up the hill.

'I'm sorry,' I said to the driver and then turned to the engineer and said through clenched teeth, 'Get it out of here and back down the hill to the lay-by. This man has got to get to Swansea urgently!'

'Can't now,' he said shortly. 'She's swung round too far!'

By now Mr Barnes had joined us and he was surprisingly calm.

'Just have to push then,' he exclaimed and took off his immaculate jacket and began to roll up his shirt sleeves.

The driver and Mr Barnes did actually try to push the lorry while the engineer looked at them in amazement. There was something about his attitude as he just stood there (although indeed, Mr Barnes and the driver were about as effective as a pair of pygmy shrews trying to push an elephant) that made me mad. I turned round and stormed back to the cottage and rang the Council.

'I do not care,' I told my friend of the day before, 'if you have to blow that lorry up and spread you know what from one end of the valley to the other, but I WANT IT MOVED NOW!'

'Can you tell the engineer to come in and have a word with me?' he said, and his voice was trembling, with laughter or fear I knew not and cared less.

His instructions to the engineer were quite explicit. The lorry was to be extracted from my gate, immediately.

The next half hour was full of shrieks, curses and heaving and the complaints of irate travellers blocked from going either up or down the lane. The idea was to run the lorry down the incline from my gateway and swing its nose around at the critical moment so that it could be re-aligned to run backwards down the hill. Of course they missed the critical moment and the lorry rested implacably across the lane. Mr Barnes was still blocked in and the traffic jam was growing.

The only solution then, was to take the gate off the field opposite and yank the wretched muck lorry into it. It stuck firmly in the mud as its front wheels were pulled into the field and its back-end still rested on the lane. However, there was now just enough room for Mr Barnes to scrape his car past and, as I'd made a quick call to his secretary to warn her to hold the fort at his function, that problem was solved.

'But we can't leave her there all night,' came the pathetic cry from the driver of the lorry as he got down from his cab and came to see where his beloved had rested. 'Someone'll bang into her!'

I didn't see how they slewed the rear of the lorry round so that it did finally get all four wheels on the lane and could be rolled backwards to the bridge. I'd gone inside by then to beat my head against the wall and get some tea for the heaving, striving gang. By the time I'd joined them at the bottom of the lane, it was clear they were in trouble once again. Somehow another critical moment had been missed and instead of the lorry being in the lay-by, it was firmly athwart the narrow bridge, where a couple of forestry workers were loudly demanding to know how they were expected to get their van past.

There was nothing for it but to call on the only reliable help I've ever found here when it comes to an all out crisis.

'Thank heaven you're there, Myrddin,' I gasped when he picked up the phone. 'Do you think your tractor could haul a lorry up a hill?'

'Give it a try,' said Myrddin who enjoys a challenge.

He appeared, just as I got back to the bridge and the disconsolate gang. A loud cheer greeted him and his big orange tractor.

'You'll never manage it with that!' said the engineer, who by now had been voted Public Enemy Number One. 'Beside which, it's nearly knocking off time.'

He moved fast, I'll say that for him, as we all advanced on him in a body; all except the little driver of the lorry that is, who was so far gone in despair that I found myself patting him on the shoulder to stop what looked suspiciously like tears.

Myrddin ignored everything except the job in hand. He clamped chains onto the front of the lorry, attached them to his tractor and motioned the driver and his mate to get up into their cab. As they moved off, I saw the driver smiling fit to burst and the forestry workers cheered wildly and broke into song. The engineer and his men hopped into their Landrover, the forestry van revved up and soon they'd all disappeared up the lane and I was alone. The river sang quietly below the bridge and a woodpigeon cooed in the pine trees.

They made it to the top of the hill and I heard later that the

Landrover then took over the job of towing the lorry back to base. Sadly, the driver got carried away and overtook the Landrover and ended up with concussion and shock when the lorry landed up in the hedge. As for Mr Barnes, when I rang him next day to apologize and to make sure he at least had survived unscathed, he burst out laughing and said: 'Believe it or not, I enjoyed it. I'd forgotten what real life was like out there in the country!'

But it isn't like that of course. Well, not all the time anyway.

=Chapter 18=

'That should do it,' said Bertie, standing back and eyeing Blossom, who was rootling around in the straw, quite unperturbed by the injection he'd just given her. 'Tricky thing giving pigs anaesthetic unless you know their exact weight. The fat tends to absorb it.'

We looked at each other uneasily. The last time Bertie had anaesthetized a pig for me, it had been to see it out of this world, and that occasion had had some long-term results which neither of us wanted to remember. Now, he'd come to ring Blossom's nose and had reluctantly agreed to put her out while he did so.

It had been my idea. Ringing Blossom's nose was a pretty horrendous business and, even if I'd been assured again and again that she probably didn't feel too much on a nose which she could use to throw enormous rocks around with, I still believed that her screams were of pure agony and not of rage. Each time I had suffered dreadfully with her and Bertie usually had more trouble reviving me

181

than Blossom. This time I'd insisted on her being anaesthetized and wanted to record the whole event, so Bertie was spelling it all out most carefully for me.

In order to give Blossom and ourselves more room to manoeuvre, I'd brought her into the big stone stable, with its wooden manger and hay racks, and Bertie had found the manger useful for parking his son Joshua in.

Looking at this tiny facsimile of his father, I found it hard to believe that so much time had passed since Joshua was merely an inconvenient bump for Sara to get behind the steering wheel of their Mini, on that snowy, dramatic night a few years ago. And I could have sworn that it was only yesterday that Myrddin and I had rushed one of my animals to the Ellis's new farmhouse and surgery, when Joshua was a few months old. While we were waiting for Bertie to finish his examination and treatment, Sara took us into the big spacious kitchen, a world away from their old shabby one, with its tiled floor and brand new Aga and pot plants flowering away like anything on the wide windowsills. She made us coffee, with Joshua perched on her hip like a small koala bear. He was an exquisite child, neat and beautifully formed. As we were seated at the table drinking our coffee, Sara calmly slipped Joshua up her navy-blue guernsey sweater, and breast-fed him. It was all done very discreetly, with none of that dreadful earth-mother flamboyance which always makes me cringe; not like the woman Mrs P met on a Sydney tram long, long ago, when she was enough of a girl herself to be excruciatingly embarrassed.

The Sydney trams in those days were bulbous, rattling affairs with open-sided compartments running from side to side. The conductors had to collect the fares by walking around on the running board, clinging for dear life to the brass rails which kept the passengers in. On this particular day, the tram was already crowded and Mrs P's compartment bulging frighteningly, when a large, fat, grubby lady pushed her way on board, holding an equally large, fat, grubby baby, bawling its head off. Squeezing her bulk between Mrs P and the open end of the tram, the fat lady beamed at the compartment, fished in her gaping blouse and brought forth a huge breast, which she offered to the baby. The baby was just as appalled as the rest of the passengers; it screamed with horror, writhing and kicking its way out of reach of the proffered sustenance. Its mother aimed a slap at it and, at that precise moment, the conductor, young and obviously not too good at keeping his balance yet, put his head into the compartment and demanded the fares.

The fat lady, struggling mightily and convulsively with her child, addressed it severely. 'Here!' she said, 'if you don't want this titty, I'll give it to the conductor!' And she thrust the vast orb straight towards the poor man who pulled back in terror, and fell off the tram. Fortunately, it had just slowed down for a stop.

The scene in Sara Ellis's kitchen was a world away from that Hogarthian nightmare. Instead, in the soft light of the converted oil lamp hanging over the table, it was more like something from a de la Tour painting and I sat watching Sara and her child with a lump in my throat. It was only when I turned and saw Myrddin going a deep puce colour with embarrassment and trying desperately to sink through a crack in the floor, that I began making a move towards the door and suggested brightly that we go and see if Bertie was finished.

Now, here was Joshua, grown into a sturdy little chap, aping his father's nonchalance and big enough to accompany him on his rounds sometimes. He was, nevertheless, still small enough to be knocked flying by an irate pig, so Bertie had lifted him up and stood him in the manger.

'It'll take her a while to go out completely,' said Bertie gesturing towards Blossom. 'Let's go and have a look at that goat,' and he hoisted Joshua out of the manger and we went off to see a rash Dolores had developed on her udder.

We came back to the stable about ten minutes later and found Blossom lying peacefully on her side.

'Good!' exclaimed Bertie, 'now I'll just get the ring and . . . Oh Damn!'

Blossom had sprung to her feet as soon as he touched her and let out a piercing shriek. Hastily Bertie threw his son back up into the manger.

'Well, Jeanine,' said Bertie in a falsely clear voice, 'I'll just have to give her a bit more I think. As I said before,' eyeing the tape recorder and its spools going around relentlessly, 'it's difficult to assess the exact dose for a pig. Now, can you just hold her steady for me.'

'I'll try,' I panted, as I struggled to grasp Blossom with one hand

and hold the microphone with the other and juggle the tape recorder round on my shoulder. 'I'll have to put this thing in the manger with Joshua I think and hang the mike up on the hay rack.' I glanced at Joshua who was looking a bit wobbly round the mouth as he gazed with huge eyes at Blossom galloping round the stable. I set off in pursuit.

'I'm surprised she's so active,' I cried to Bertie as I hung on to my pig. 'I mean you'd think the anaesthetic would have at least slowed her down.'

Bertie looked at me wildly as he strove to find a vein somewhere on Blossom.

'There now! That should put her right out!' he declared triumphantly at last. 'We'll just leave her again for a while to make quite sure.'

When we came back again, Blossom was snoring loudly. Bertie put his hand on her back. Blossom leapt, grunting, to her feet and bashed him hard. This time I hurled Joshua back into the manger. He immediately began to bellow furiously.

'Shut up, Joshua!' yelled his father. 'I forgot, he doesn't like pigs actually!' I gave him a very sour look.

It took us a long while to catch Blossom this time, enraged as she was by Joshua's shrieking. Finally I ran her to ground and sat on her head while Bertie advanced with his syringe.

'She can have just the one more and that's that!' he puffed. 'I daren't give her any more. Why the hell isn't she out? She must be a lot fatter than she looks!'

'Well if she's not out by the time we come back this time, we'll just have to ring her as usual,' I said. 'Let's go into the house and have some coffee and get Joshua something to cheer him up.'

We should have left Joshua behind with Mrs P, but neither of us imagined for a moment, that this time Blossom wouldn't be dreaming happily of whatever it is pigs dream of. She *looked* as if she was dreaming all right, but the moment she heard us, she exploded with wrath. We barely had time to fling Joshua back into his perch and then began the race to end all races as Bertie and I flung ourselves around the stable until we both landed, gasping, on top of Blossom. Somewhere, in passing, I switched the tape recorder on again.

Usually when we rung her, we had little trouble getting a halter round Blossom's nose and, although she screeched shockingly before the ring went in, by then I was busily rubbing her stomach, which always calmed her and, if Bertie was fairly nimble, he got the ring in smartly. This time, no amount of tummy rubbing or soothing words had any effect. Blossom fought us every inch of the way, keeping up a high pitched noise level which nearly sent us over the edge, and Joshua matched her shriek for shriek. By now both of us had forgotten the tape recorder, faithfully whirring around, and the microphone dangling from the hay rack. Our language, to both pig and child, was truly appalling.

When we had finally finished, Bertie and I leapt for the door, collecting Joshua on the way, and stood peering through the window as Blossom oinked her way savagely round the stable. Joshua hiccuped miserably beside us. At long last Blossom swayed and sighed and sank to her knees. She laid her head on the straw and, still grumbling, settled herself and finally slept.

Bertie and I breathed again. 'She'll sleep for quite a while now,' said Bertie. 'I was hoping to avoid that though, which is why I gave her such a small dose to begin with. You can understand why I wasn't too happy about the whole business in the first place. Anyway, leave her alone and just check on her every hour. I'd better get this little chap home to his mother.'

When they had gone, Joshua quietly sobbing to himself in his little car seat and Bertie shaking his head as if he couldn't hear too well, I tiptoed back to the stable to collect the tape recorder. Blossom still lay asleep, only her ears twitching fitfully from time to time.

As I closed the door behind me, I heard her grunt crossly in her sleep.

'Have you checked on Blossom again?' said Mrs P an hour later.

'Er . . . no.' I replied slowly, 'I've been listening to that tape. It's quite unusable of course. Terribly over-recorded and I didn't know I knew language like that!'

'Oh, you do,' said my mother serenely, 'though where you get it from I've no idea!'

I looked at her in amazement. I knew where I got it from all right, but there was no use saying so.

'Right!' I said, 'I'd better get out and check on this pig.'

Carefully, I lifted the latch of the rickety old door of the stable. It creaked and stuck and I held my breath as I pushed it hard. I peered round the door fearfully, but Blossom still lay as I'd left her, her sides going up and down rhythmically and one of her big ears covering her face. She was once again my dear old amiable lady. I felt my heart swell with pity for her and this awful business of having to get a ring in her nose to stop her joyful rooting of enormous holes in the hillside.

I'd hoped, last time she'd lost her ring, that she might just have forgotten about digging, but soon the field had begun to look like a battleground and, one evening, it was used by the army in rather that way.

Mrs P and I had been sitting peacefully by the fire at dusk, when I noticed, through the window, a figure creeping furtively across the yard. I shot out of the door and shouted, 'Oi! Where do you think you're going?'

The figure straightened and came towards me, neat and trim in his battledress. He saluted and apologized. Still smarting from my previous encounter with the army, I folded my arms and frowned at him while he explained that he was on an exercise and at all costs must avoid being spotted by the 'enemy' who were probably, at this very minute, training their field glasses on him from across the valley.

'Well, as you're already half way across, you'd better keep going,' I said.

'Does that mean all of us?' he asked.

'*All of you*!' I cried. 'Where?'

He gave a low whistle. Suddenly, the whole of Blossom's field was alive with helmeted heads draped with bits of foliage. My trespasser gave another low whistle and, as one man, what looked like a whole platoon rose from the dug-outs Blossom had so thoughtfully provided for them. Another whistle and they began to run, doubled up, down the hill. I stood there in shock as they reached the yard and began crouching and creeping around the garden hedge towards the gate. When they were almost there, a dreadful, blood-curdling yell from the field opposite sent them all flat on their faces.

It was my neighbour, Cliff, violently abusing his sheep-dog, but it took me a while to tell the soldiers that, because I was doubled up myself by then, laughing like a maniac. Nevertheless, I'd thought, as the fugitives got to their feet and drifted, one by one, down the lane, Blossom's digging would have to stop. Unless I hired her out to the army.

'Poor old Blossom,' I said to her now, and laid my hand lovingly on her back.

The effect was so instantaneous, so stunningly violent that to this day I don't know how I got out of that stable alive. Blossom's great body leapt up in one almighty bound and she landed running. Every wild, human-hating ancestor, was there in her face, contorted now with rage and intent on murder. I made it to the door, which mercifully didn't stick this time, and slammed it shut as she came hurtling into it behind me.

Frantically fumbling with the latch to secure it, I heard Blossom back off, grunting viciously, and then the sound of her body thudding into the door again. Too many onslaughts like that and the door didn't stand a chance.

I took a quick look through the window and saw Blossom lining up for another charge, only pausing to savage the wooden struts of the manger in passing.

'Food,' I thought, 'she'll listen to food. If I can get her out of there and onto the field, at least she can bash about without hurting herself or the buildings.'

It worked, up to a point. By the time I'd got the bucket of nuts jangling and rattling outside the door, Blossom had calmed down enough to listen. I opened the door and showed her the bucket.

For a moment, her enormous greed penetrated the fog in her brain and Blossom trotted rapidly towards me. I ran, faster than I thought I could, straight through the open gate of the field, and flung the bucket towards Blossom who'd just begun that awful hysterical screaming again. She picked the bucket up, tossed it in the air and tried to tear it to bits. That gave me time to get out of the gate and close it quickly before she remembered she was going to kill me and charged again at speed.

It was midnight before she stopped ramping and raving and rattling the gate, while I prayed that it would hold and Mrs P kept saying over and over again, 'You realize you've probably damaged her brain or something, don't you!'

'That pig,' I told her, 'is raving drunk! And she's not a nice drunk either!'

At last I heard Blossom stamping her way crossly into her shelter, which lies just above the garden hedge. After a while, the sound of loud snoring blasted the calm night air.

In the morning, there was absolutely no sign of Blossom at all. I was torn by desire to make sure she was all right and by the fear that her manic fit would be roused up again if I went near her.

I stood by the gate with the eternal bucket of nuts and rattled it tentatively. I was answered by a low grunt from somewhere deep inside the shelter and gradually the tip of Blossom's snout poked out. The rest of her head began to follow and I rattled the bucket invitingly. A groan answered me and the rest of Blossom emerged slowly into the sunlight. She stood there, shaking her ears very gently as if arranging them to block out the light, and began to walk carefully towards me. She stepped across the grass as if it would explode on impact, picking her feet up and placing them down again with infinite care.

A couple of sparrows began squabbling shrilly in the thorn tree above her head and Blossom shuddered. Very, very slowly she lifted her great head and looked up at the birds with loathing.

'Talk about "alka-seltzer don't fizz"!' I said to Mrs P. 'Blossom has got a massive hangover!'

'I'm not surprised,' she retorted fiercely, and spent the rest of the day making up little milk puddings, laced with a selection of nerve-soothing herbs, which Blossom accepted from her languidly, before retiring once more to her shelter to sleep it off.

Mrs P's ministrations, however, had little effect on calming down Winston. He was going through a dreadful identity crisis and nothing she could feed him or say to him was of any use.

As he'd grown older, Winston had turned into a very randy little dog. The end of his black-button nose had gone quite pink from being constantly poked where it wasn't wanted, and the bitches were tired of having to keep their behinds firmly clamped to the ground when he was about. Mrs P tut-tutted at him, but was secretly quite proud of her little Don Juan, and when a lady rang to ask if she could bring her chihuahua bitch to be served by Winston, she was delighted. Especially as there was a healthy fee involved.

'At least,' she crowed, 'he'll be earning his keep unlike some other dogs around her!'

'You may live to regret it,' I told her. 'Once he gets the idea properly, he'll want more of the same you know!'

'I'll cross that bridge,' said Mrs P firmly, 'when I come to it.'

The little bitch, when she arrived, was enchanting. She was a pale cream and of a charming disposition. She sidled winsomely up to Winston and prepared, happily, to be ravished.

Winston took one look at her, went down on his belly, crawled along the floor to Mrs P's feet and looked up at her piteously.

'Don't be silly,' said Mrs P. 'This is what you've been waiting for!'

Winston gazed at her hopelessly and crawled under a chair.

I smiled sickeningly at the bitch's owner. 'It's just that he's never seen a chihuahua before,' I explained, 'well not since he was only a puppy. He thinks he's a whippet.'

Winston's little face peered from under the chair and Mrs P hauled him right out. She marched over to the big mirror with him.

'You,' she told him, 'are a chihuahua!'

Winston snarled at his mirror-image. Chihuahuas, his expression said, were revolting!

We tried everything. We left them alone and peered at the happy couple through a crack in the door. Winston cringed as far away from the little bitch as he could, and piddled miserably on the floor.

I couldn't believe it. This was the second time a dog had let me down like this. Mine, because Mrs P had suddenly pretended he was nothing to do with her.

It took Winston a good week to get over it. He never left Mrs P's side and she told me he was crying pathetically in his sleep.

Mrs P was disgusted. 'He's like a lot of men,' she said indignantly. 'All talk and no performance!' And the next time Winston tormented one of the whippets she picked him up and slapped him.

'If you don't do it when you're supposed to,' she said, 'you don't do it at all!'

But Winston, time healing the memory of his failure, began once again to waltz around on his hind-legs after the bitches and finally, tired of the snapping and snarling, and with Mrs P's love of her dog diminishing daily, I got Bertie to give Winston an injection to cool him off.

'What is it?' I asked him as he brandished the syringe.

Bertie smiled evilly. 'Best Fairy Liquid,' he declared.

Whatever it was, it worked very well for a while and Winston's nose got quite black again.

Chapter 19

She is my Lily Maid of Astolat, my Lily the Pink, my Diamond Lil', my Jersey Lily, and most of all my big, beautiful, golden Tiger Lily. We met, quite by chance, in a warm, dimly-lit barn in Roxburgh-shire. She was only the size of a small, striped football and we were both in a hurry; she to get to the milk queue first and I to get away from the temptation of nine lurcher pups, but when she blundered into my legs and I stooped to put her back on her fat floppy paws, the outcome of our meeting was already fairly certain.

I knew her father of course, in fact had known him when he was only a young dog himself, standing, on that bright cold morning, outside the cottage in the Cotswolds, the burnt orange and black of his coat making a bright statement against the grey stone walls, and his lean, muscled body leaping around his owner as she greeted us with offers of warmth and breakfast and the use of a telephone.

The Artist and I had left London when it was still dark, in spite of forecasts of heavy snow. We ignored them because this was to be the big day on which he, at least, would leave London for good and settle at last on the farm in Wales while I continued to commute back and forth. On the roof-rack of the old Beetle (FFU herself) were lots of canvases and inside, the car was stacked solid with his belongings, including a big cauldron in which, for some reason, he intended to cook his meals over the open fire. Not until one of us was actually settled on the farm did we feel that it was really ours and the long weeks of negotiating for it quite behind us, so we scoffed at the forecasts and set out.

The car radio, however, was definite. Very heavy snow was predicted for South Wales and if it wasn't there already, it soon would be.

'Perhaps we should stop and ring the Summers,' I suggested at last. 'I mean, if we can't actually get to the farm at all it's a long way to come back in a day. Why don't we stop off at Leesa's in the Cotswolds and then at least we'll only be half-way.' And so we had turned off and gone to Hampnett and the converted school-house which Leesa Sandys-Lumsdaine shared with a pack of greyhounds and lurchers. There she painted her pictures of horses and dogs and other animals, like the pet pig which used to trot behind her own horse when she went riding.

As soon as we entered the cottage, where Leesa's mother was calmly sewing rabbit skins together to make lures for training the dogs, we rang Imogen Summers, but it was her daughter Rosemary who answered.

'Snow?' she said, surprised, 'no, nothing of the kind. Well, perhaps, now I look out of the window, there are one or two flakes falling, but it's not settling. I don't think you'll have any trouble getting through.'

'Nevertheless,' I said to Leesa, 'we really ought to make tracks quite quickly, just in case.'

'Oh, stay and have some bacon and eggs,' said Leesa and, unable to resist the heavenly smells coming from the frying-pan, we did, while the dogs sat around and politely and hopefully thumped their tails.

'That is Tiger,' said Leesa between mouthfuls, pointing to the splendid dog we'd seen outside. 'I'm not sure exactly what his breeding is, but the gypsies I bought him from told me he was half

Doberman and half greyhound. They were camped not far away and a friend of mine and I were out hacking when we saw this poor thin bitch trying to feed her pups. So we dug in our pockets and put our money together and bought as many of the pups as we could, to rescue the poor things and give the bitch a chance. I kept Tiger here myself. Isn't he terrific? Do you know, that dog can jump his way out of a tennis court?' Tiger, knowing we were talking about him, sat back on his haunches and grinned at us.

We left soon after that and I didn't see Tiger again till he was an old dog and Leesa had long since moved to Scotland. But he and the bright cheerful cottage, full of warmth and welcome and colour, were about the only good things about that week-end.

The Artist and I had driven on to Wales and sure enough there wasn't a sign of snow. The A40 wound through Brecon and on past Sennybridge and still the countryside was clear. As we went further, we did notice a few patches of white lying in the fields but, as we took the turning to go across the mountain, the road ahead was a solid, unblemished white.

'Do you think we ought to go round the other way?' I asked as the Beetle started slithering and sliding.

'No, it's much longer and I don't think this is too bad. We'll just take it slowly,' he said.

It was fine for a few miles. The snow was new and not very deep as yet and the dear old Beetle churned its way round the bends and never offered to stop; not, that is, till we reached the half-way mark on a long hill, when it tried and tried but finally gave up with a sad little cough and stopped dead. Somehow, by jumping up and down on the bumper and heaving and pushing, we got it turned round and set off back to the main road. We were about a quarter of a mile from the village when, with a roar and a scraping, along came a snow-plough. Laughing triumphantly, the Artist let it pass and then swung the Beetle around and tucked in behind it.

We sailed along in its wake, so busy congratulating ourselves on our incredibly perfect timing, that we almost banged into the snow-plough as it came to a sudden halt just before the bridge over the infant River Usk. Horrified, we watched it ease forward, reverse onto the verge and head back towards us. Beyond it lay the mountain road, very deep in snow and quite untouched. As the snow-plough drew alongside, I leapt out of the car and hailed the driver.

'Why are you stopping?' I cried. 'There's miles to go yet!'

'Not our concern,' chirped the driver. 'This is a Powys snow-plough and over the bridge it's Dyfed! It's up to them to clear the next bit.'

Sorrowfully we turned the Beetle round and followed them back to the A40 again and went the long way round. A few claustrophobic farmers had been busy and our road was clear. We called in on Rosemary as we passed her farm.

'Funny thing that,' she said, 'it began to snow heavily just after you rang off, but I didn't know the number to get in touch with you!'

The Artist didn't stay at the farm that week-end after all. His canvases had been soaked through, his morale was low and the snow was followed by one of the worst frosts I've ever known. Using the excuse of it being far too dangerous for me to drive on the frozen roads alone, he took me back to London and left his Great Remove till the weather was kinder.

It was blazing with kindness when I saw Tiger and Leesa again. It was the beginning of summer and, in the big sycamore tree overhanging another converted school house, the sound of bees was deafening.

I'd been too tired the night before to recognize Tiger in the barking mob which met us. Too much had happened that day and my friend Gay and I were just plain relieved to see the lights winking ahead and to know, as we drew up, that we wouldn't be spending the night stranded in some forestry glade or worse. Gay had been doing the driving and I the navigating, head down consulting maps and the wads of typed instructions which would hopefully land us up at all sorts of out of the way places in Cumbria and Scotland. So far they had, but Gay and I had had a nasty few miles of it on the final leg to Leesa's.

We'd set off from Wales a couple of days before, on a Sunday morning, and had spun round the empty roads which took us past the Brecon Beacons glowing in the early light, a deserted Hereford, and finally out on to a motorway gloriously free of traffic, till we caught up with the erratic stream of cars pounding up to Blackpool or the Lakes for the day. We'd got enough of a start, however, to be settling into our Bed and Breakfast place at Crook by midday. We were still fresh enough to spend the afternoon with a lady dry-stone-waller high up on a fell scoured by the wind, and to go on and claim another victim in the evening.

I don't know what magic is in the air in Cumbria. Gay and I looked out on spectacular scenery every day of our lives, so the views were not the only reason for the way we felt; full of that anticipation of something quite amazing just about to happen. It gave us enough adrenalin anyway, to cope with a breathless day of talking to people on the National Trust farms; of a wonderful encounter with two grand ladies and their russet-coloured goats and fell ponies, and still to rush on to Kirkby Stephen just in time to catch the sight of a brilliant flock of free-flying macaws come skimming across the evening fields.

It was eight o'clock before we could tear ourselves away from the lovely old house surrounded by its cacophony of parrots roosting in the trees, and head for Scotland. We were making for Leesa's new home, several miles into the forestry beyond Roberton. Leesa had sent a detailed map and beside Roberton she had written, 'Look carefully, or you might miss it!'

'Roberton looks all right,' I said to Gay as I peered at the map by torchlight, 'what doesn't look so inviting is this track she's drawn in, that leads forever over nowhere.'

'We'll manage,' said Gay cheerfully, still full of energy in spite of the long hard drive we'd just done in record time. 'I think this is Roberton, that cottage over there!'

'Look for a left-hand turn then!' I said quickly as we shot past. 'Only I think that was it back there!'

We back-tracked a few times, found what we fervently hoped was the proper turning and didn't lose our confidence in our choice until the road crossed a small bridge and began to narrow ominously. Every now and again there was a bulge in the road and a warning notice saying 'No Parking! Passing Place Only'.

'Now I know what people must feel like when they visit us,' I said to Gay, as the road wound on and on and promised nothing for the immediate future but a dark mass of forestry plantation ahead.

'A light!' cried Gay, 'no, sorry, it's just another car.'

'Well at least it means someone else uses the road,' I said hopefully.

I shone the torch down on Leesa's map again. 'It says here that she's just by a phone box! I can't imagine a bright red phone box stuck out here, can you?'

'No, but I can *see* one and that, if I'm not mistaken, is Leesa's house!' said Gay.

And there, sure enough, was the steep roof of what had once been the village school house and emerging from it, as lights blazed on, came a great pack of longdogs, and Leesa herself, this time offering dinner instead of breakfast.

'But it will have to be fairly quick,' she said, bringing her voice down to normal after shouting above the dogs who stopped their barking, yawned and sauntered back to the house. 'They're expecting you down at the farm!' For Gay and I were booked in to stay at a farm further down the road and, if Leesa was tolerably happy to stay up for us, the farmer's wife was anxious to get to bed.

So, in all the rush, I didn't notice Tiger particularly. I did remember, however, that Leesa had told me she had a litter of nine lurcher pups out in the barn. As we tiptoed round the farmhouse and found our rooms, I said to Gay, 'To-morrow, whatever you do, don't let me go near Leesa's pups! Mrs P said that if I get any more dogs, she's off back to Australia. I never could resist a lurcher pup!'

'Don't worry,' Gay yawned. 'I'll keep an eye on you. I'm not going to have Mrs P pointing the bone at me for encouraging you.'

It was easy to avoid going near the pups the next morning. Gay took the younger dogs off for a walk in the forestry while I interviewed Leesa about such things as the hazards of being a country artist (she once found a large, fat cow trying to climb into the back of her Volvo); about the local custom of breeking the ewes, which they didn't want served by the ram, by sewing bright patches of cloth to their behinds (and being offcuts from the mills, those patches could by anything from flowered cottons to tartans); about the dubious pleasure of having a family of red mice living in her car and of the way some of them rolled out when she was driving along or, even worse, of the time they chewed up a pile of tenners she'd left in the glove box, and used them for nesting. While we talked, Tiger, now an old dog but with the same wicked, curling grin, lay at our feet and thumped his tail as if agreeing with everything Leesa said; bantams crowed or clucked around us and, in the tree above our heads, a large party of wild bees kept up a steady roaring buzz.

After lunch Leesa had arranged a heavy schedule for us, which entailed driving off to see a friend of hers who bred Clydesdales (and turned out to be an old school friend of Gerald Summers) who took me for a long, wonderful ride around his farm on the flat hay-cart pulled by one of the great horses.

Our next call was to Lilliesleaf to the local 'goose-sexer' who was a charming old man, not at all sure why I was standing there with a microphone as he exhorted his pet Chinese goose, 'Will ye no lay an eggie for me dear!' and she hissed back at him derisively. He'd had her for twenty years, since he hatched her out in his airing cupboard. As for the rest of the geese, I never did find out exactly how to sex them. Well, I already knew the principle, but it takes a very sure and gentle hand to discover if a goose has the right equipment to make it a gander and it's something most people would rather not do. Which is why people for miles around, including Leesa, brought their young geese all the way to Lilliesleaf for him to 'sex'.

We did find out, however, that he'd been breeding Light Sussex hens for many many years and two of his hens, hatched out from some eggs we begged from him, are still being chased out of my garden to this day.

Gay and I got back to Leesa's full of our adventures, the car laden with eggs and goose feathers and our guards completely down. We sat over dinner, watching the light greening into dusk. All around us on chairs and on the floor, lurchers and greyhounds slumbered. Outside, Leesa's bantams were having a last crow before making their way off to roost, the wild bees had ceased their humming and, across the little road, a farmer was making a call in the red phone box, his collie waiting patiently outside.

'We'd best be going back to the farm soon,' I yawned. 'Long day tomorrow. We've got to see a breeder of collies over near The Hermitage, get to Dumfriesshire to see a spinning-wheel maker and then find our way back to Cumbria for the Herdwick Ram Show the day after. I'm going to meet an old shepherd there who used to work for Beatrix Potter!'

'Oh! But you must see the pups before you go,' said Leesa.

'They're Tiger's you know. Tiger's and Cuckoo's. They've been in love for years, but she has had to go to proper greyhounds to be mated and they've loved in vain!'

I nodded. Cuckoo was a very grand greyhound who's real, registered name was Happy Wanderer. After a successful career on the track, she'd won the Scottish Grand National, the big coursing event and one of her sons was to win the Waterloo Cup the next year.

'They're both getting on now though,' said Leesa, 'so I thought, just this last time, Cuckoo could have a litter by Tiger and she went and had nine! They're just five weeks old but Cuckoo's a bit tired of them now. Actually, it's time she fed them.'

'You can *look* at them then,' Gay whispered to me, 'but that's all!'

As we moved towards the back door, most of the pack got to their feet, stretched, and padded quietly after us. Cuckoo, a beautiful fawn bitch with a dreamy expression and a slight tilt to her nose, went ahead of us and waited for Leesa to open the barn door. A few bantams, coming in late, joined us and a self-important muscovy duck waddled from behind a bush with a flurry of sleepy ducklings after her.

Inside the barn there was nothing to be seen but bales of straw and a couple of tea chests turned on their sides. Cuckoo went over to the nearest tea chest and nosed inside. Instantly pups flew out in all directions, bowling and tumbling over each other, and heading the rush was the smallest one of all. It bumped into my legs, I bent to steady it, it flung itself up and licked my nose and then raced over to argue the last of Cuckoo's teats from a pup twice her size.

We stood for a moment, watching the pups as, their meal suddenly curtailed by Cuckoo, they rolled and fought and racketed about. Leading every wickedness, always on top of the heap was the small pup, the only one to have inherited Cuckoo's turned-up nose but, like the rest of the pups, a warm golden brindle like her father.

At last, worn out, the pups staggered back into their tea chests and fell asleep, one on top of the other. Leesa closed the barn door and led us back to the car.

'How much do you reckon to get for the pups?' I asked her casually as we walked and I swear that at that moment it was just a casual enquiry.

'I'm asking £25,' said Leesa. 'I'm keeping a couple for myself and frankly I'm more interested in who has them than in what I get for them. Anyway, it's early days yet. They're not due to leave their mother completely for another three weeks.'

My bed at the farm was warm and deep and comfortable. Nevertheless, I didn't get much sleep that night. Whenever I drifted off, the smallest pup came floating into vision and with her came dear old Tiger, not as he was now, but as I'd first known him, young and gleaming and full of life.

We left very early the next morning, Gay and I, so early that I was surprised to see a light on in Leesa's little house.

'Stop the car!' I told Gay suddenly. 'I just want to give Leesa something.'

'Well don't be long,' she said as she stamped on the brake. 'We've got a fair few miles to go today.'

Leesa was in her kitchen, still in her dressing gown and absolutely amazed to see me. She was even more amazed when I handed her a cheque.

'What's this for?' she asked.

'The pup with the turned-up nose,' I said.

'But I thought . . . I mean, you said no more dogs! What about your mother?'

'I've got three weeks to worry about that,' I said.

The car horn blared suddenly and I leapt for the door. All the dogs leapt with me and Gay and I drove off with their barking still ringing in our ears and a last vision of Leesa in her dressing gown waving the cheque at me, in the rear vision mirror.

'I hope,' said Gay, 'that you make it quite clear to Mrs P that I had nothing to do with this! If I'd known what you were up to I'd never have stopped the car!'

'Mrs P will know that nothing and no one could have stopped me,' I assured her. 'But honestly, I didn't have any idea what I was going to do till I saw that light on in Leesa's place. Feel like coming back to collect the pup when the time comes?'

Gay grinned happily. 'Now there's a thought,' she said.

As it turned out, we didn't have to go back to Scotland. It was three weeks, almost to the day, when Leesa rang to say she had to come to London because she had an exhibition of her paintings at the Tryon Gallery. If I could meet her there, she'd bring the pup with her.

Once again, Gay and I set off. We left the car just off Kensington High Street in the car park, and that's why we missed all the razzmatazz of the drinks party to celebrate the opening of the exhibition. Back on my old stamping ground I had to call in on friends and go and buy lots of things from the patisserie and show Gay the little streets I'd once known so well.

We finally made it to the gallery when everyone had gone to lunch, but at least we were able to wander about in peace and look at Leesa's paintings without a crowd. And looking at them, it seemed as if we were indeed back in Scotland again. There was the bright red phone box with the farmer and his collie, both inside this time, while outside, Leesa's muscovy duck, attended by her ducklings, was pecking furiously at one of the lower panes of glass; there was her greyhound bitch, Strippit, stealing Leesa's bra off the clothes line; there was a litter of piglets playing football with a large swede; there was a fox picking up the crumbs under the bird table; there was Leesa's favourite cockerel crowing his heart out; there, in fact, was Leesa's world beaming out at us from the walls of a smart London gallery and bringing with it the scent of pines and the humming of bees. Of Leesa herself there was no sign and I had a nasty feeling she might have already left, pup and all, and gone back to Scotland.

That thought cast a slight shadow over the next half hour which I spent with John Newth in a funny little cafe across the way while Gay took herself off to have a quick look around the shops. Carefully I eyed the gallery door while John and I lunched off sandwiches and I signed a contract; a contract to write the book which John had been nagging me about for months.

Poor man, if he'd known how much longer it was to take him to get that book and how many deadlines would pass, he might not have thanked Gay so heartily for bringing me to London, nor caught the mood of euphoria ricocheting around the walls of the gallery as we returned and found Leesa and her friend, Penny Lowes.. We got their assurances that the pup had travelled beautifully and was awaiting us all at the family home in Buckinghamshire. They had left her with Leesa's mother and travelled in by train to London, so now, with the gallery closing and many of Leesa's paintings already marked by the red 'sold' stickers, we could all go back together in Gay's car.

I don't know what it was about that day, but every detail of it is vividly and gloriously clear, even the frantic search for a taxi to get us back to Kensington to pick up the car, and me sighting one several blocks away (a taxi driver once told me I had 'Regular Rider' written all over me, which is why he stopped instead of going round the corner to pick up a rich passenger from the Hilton); the meeting at last with Penny who had been Leesa's partner in setting up the now famous Lambourn Lurcher Show years before; the gallons of tea we drank in the elegant room in Buckinghamshire, where Leesa's mother was confined to the sofa with a broken leg and her whippets lay about on the special dog-beds Penny made in her Yorkshire farmhouse; the glis-glis or edible dormice which had set up their home in the airing cupboard and could be heard making their shrill little noises all over the house; the sight of my pup, worn out with the joy of meeting us and thundering about the garden on her big ungainly legs and no one minding a bit when she piddled all over Mrs Sandys-Lumsdaine's carpet. It was one of those rare days when everything was absolutely all right.

Lily slept all the way home to Wales, even when we stopped at Northleach and rounded off the perfection of the day by standing in the dusk looking at the loveliest church in the world; or so I think and never pass Northleach without going to make quite sure.

It took Lily a couple of days to win Mrs P over, but win her over she did, although she was the most difficult pup to train I've ever had. She cost me a fortune in chewed slippers and shoes, jumpers and cushions and even my reading specs. The first thing she did to one of Penny's splendid beds when it arrived Special Delivery from Yorkshire, was to piddle all over it so that it took days to get clean and dry. The only thing in the world she was afraid of was Winston and it was a long while before the stock outside could take her for granted.

In spite of her wickedness, as every day went by, Lily grew more beloved, for she had brought with her something of the lovely place in which I'd found her; something of Leesa's unique view of the world around her; some sense of joy which I cannot for the life of me put into words but which Lily seems to carry about as part of her basic equipment. She has Cuckoo's calmness and Tiger's wicked grin and I have only to see her sitting upright in her 'PPP' (Pile-

Preventing-Position as Leesa explained) with the daylight visible under her big haunches and her face with its tip-tilted nose twitching in anticipation of dinner (for Lily treats all food with deep respect in the best gourmet tradition) to get a lift of the heart and forget whatever odd sling or arrow is aiming itself in my direction. She has grown into a big dog and that in itself is a comfort on those nights when the cottage creaks and the senses freeze, wondering . . .

That winter, I blessed the whim which had made me dash in to Leesa's cottage and write that cheque. It was a dreadful time of blizzards which made every previous one seem a mere flurry of snow. Between one onslaught and the next the Blight made a brief but vicious return and when it left, three loved faces were missing. It was Lily, with her great beauty and the sheer solidness of her lying across my feet, who eased the pain. And the sight of her, digging a full six feet down through a snow drift to recover one of her precious bones, made even the blizzards seem of no account.

It took John Newth another two years to get his book from me. It was hard to go back into the past while all around me the present was making itself into instant copy and it was hard to write about the star of the place when she was missing. I'd seen her once at Sedley Sweeney's, the proud mother of a big, amiable Shire foal, called Task Force (Stanley for short) which had been born on the day the Falklands battle was won. I'd heard about her often from Sedley; about the way she'd careered off pulling the plough, just as he was demonstrating to a group of students, and he had an awful time getting her back; I'd heard how, in more willing mood, she'd carted the feed out to Sedley's sheep during the blizzards and carved great swathes through the snowdrifts and of how he rode her on long treks across the mountain. I knew she was well and happy and useful, but still I missed her dreadfully.

Chapter 20

An Indian Summer lazed across the valley. The woods, braced for autumn, relaxed, and only a keen eye could see the tinge of russet veining their deep green. Light beat back from the white walls of barns and cottage with no haze of mist to blunt its sharpness. The Fans etched a knife-edged silhouette against a clear sky and there was no shadow to deepen the long red runes which scar their sides.

Beyond the gate, the little lane wound steeply away out of sight and nothing came to ease the strain of listening. On the home field the goats sat lazily cudding against the hedge, eyes glazed, not seeing, not caring. Of the sheep, long since gone to the cool of the river, there was no sign. Ducks stood, heads turned over their backs, as they dozed beside the pond and chickens lay in companionable groups, lazily flicking dust over their backs or scraping themselves deeper into the hollows against the stone walls.

It was the geese who heard it first. Heads up, necks out, they honked the alarm and streamed aggressively towards the gate of their yard. And then there it was, the slow grinding of gears and the sight of the horse box just appearing over the hedge as it descended the hill. Distantly, from the cottage, I heard Winston yapping. The pink door opened and Mrs P came out to stand in the garden and watch the arrival. Slowly, heart thudding, hands fumbling, I went to swing the gate open and wave the Landrover and box into the yard. As it passed me and came to a halt, I heard the sounds I'd missed so much; a loud enquiring neigh, the impatient stamping of heavy hooves and a long resounding snort.

I banged the side of the horse box and a deep snuffling came from within.

Sedley Sweeney leapt from the Landrover and shook his head at me.

'I couldn't believe you wouldn't have her back,' he said.

'That horse,' I told him, 'has a charmed life and for some reason, fate has decided she will spend it with me.'

Sedley smiled. 'There's certainly something odd about it. Isn't this the second time this sort of thing has happened?'

'Yes, and I'm not risking a third time lucky. This time she's here for good, well at least till Him Up There decides otherwise.'

Many years before, when the very thought of parting with Doli was quite unbearable, Sedley had been trying very hard to buy her from me. Financially in grave trouble at the time, I almost gave in to him, but a totally unexpected cheque for almost the same sum as I could have sold her, came just in time. And now, with the situation reversed, very much the same thing had happened once again.

Sedley had rung a couple of weeks before, to say that he was selling up the Smallholder's Training Centre and that although he'd managed to find good homes for all the other animals, Doli was proving very difficult to place.

'You see,' he'd explained, 'I can't really put my hand on my heart and say she is trained to the plough, not since the time she took off with it. And she really isn't too reliable as a riding horse and not many people want to ride a draft horse anyway. She's getting on a bit too and so really as a brood mare she's not a very good proposition, especially as she's so hard to get in foal. You know she was empty again this year? So, I fear the only fate for Doli is to go to market where the meat men will bid for her, or be taken to the knackers at Bristol, or to come back to you!'

'I can't afford her Sedley,' I'd said, for in spite of being in work again, things were still very lean and Doli's purchase price, her feed for the coming winter and her nasty habit of running me up vet's bills would pare everything to the bone.

'I'll let you have her for £200 less than I paid for her!' said Sedley. 'I did quite well out of Stanley her son, so she's more or less paid for her keep.'

'That's a very good offer,' I said, 'but I'll have to do my sums very carefully. Can you give me a while to make up my mind?'

I tried not to think about Doli for the next few days, firmly telling myself, in spite of Mrs P's anguished looks when I told her of Doli's fate if I didn't have her back, that for once I had to face up to reality and stop letting my heart rule my head. I was kidding myself of course.

When the television company rang to ask me if I would do a voice-over for a series of programmes they were making for Channel 4, I thought they'd got it wrong. I'd never done a voice-over in my life before and I imagined they must have got me mixed up with someone else. But no, it seemed the producer had been very undecided about who he wanted for this particular job until, with everyone nagging at him that the programmes were nearly complete and he'd better make up his mind fast, he'd heard *A Small Country Living* on his car radio. He made an instant decision and told his secretary to get in touch with me at once. As she was speaking, I was doing sums rapidly in my head. The fee they were offering was healthy enough to pay for Doli, her winter feed and a few excursions by Bertie.

As the great feathered feet plodded down the ramp of the horse box and Doli finally stood on the yard letting out a great indignant neigh and pointing her head meaningfully in the direction of her old stable, it seemed impossible to imagine she'd ever been away.

I patted her great neck. 'Hullo you old sod!' I said.

As Mrs P said later, 'There wasn't a dry eye in the house.'